fP

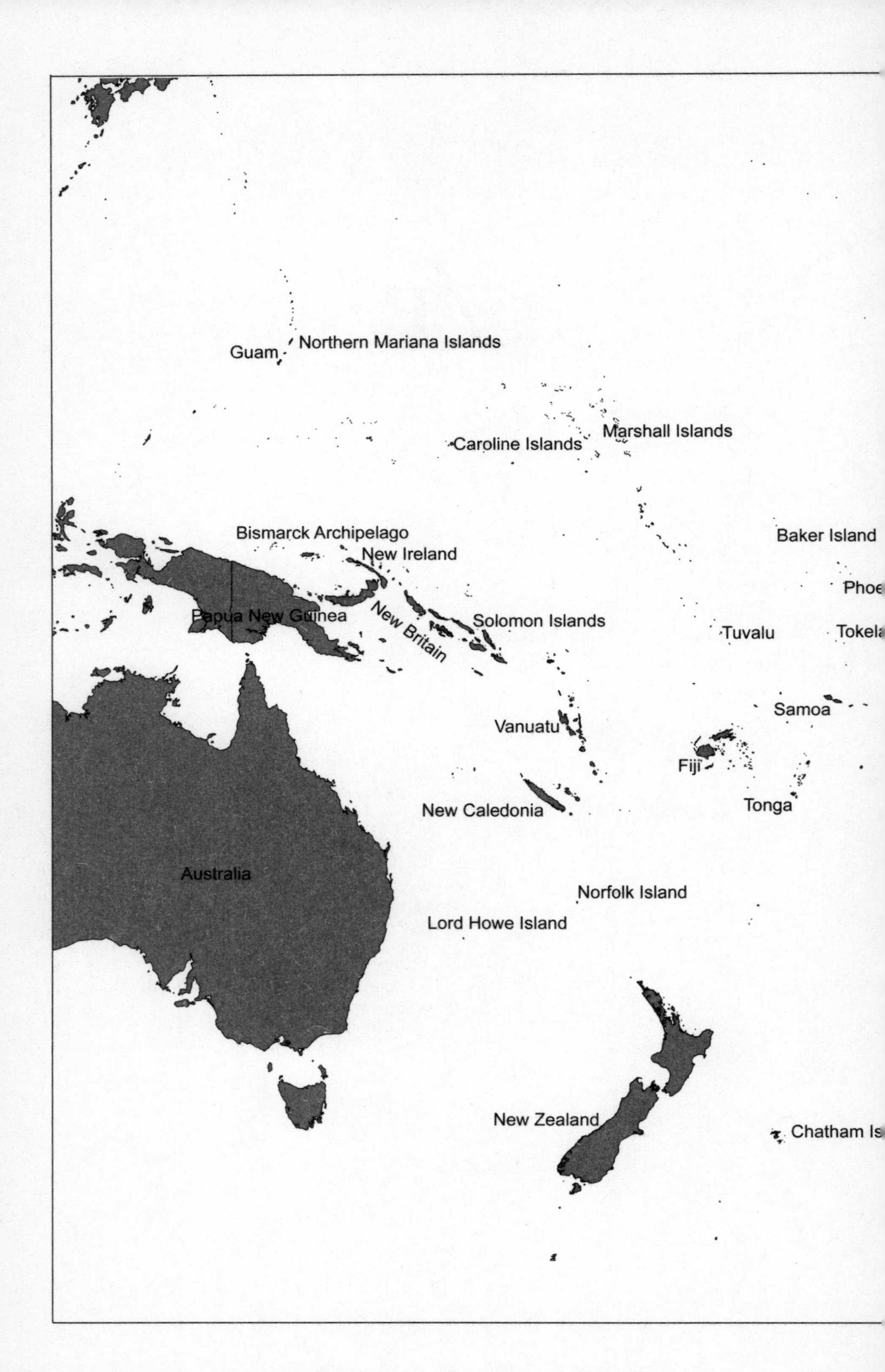

Northern Mariana Islands
Guam
Caroline Islands
Marshall Islands
Bismarck Archipelago
New Ireland
Baker Island
Papua New Guinea
New Britain
Solomon Islands
Tuvalu
Samoa
Vanuatu
Fiji
New Caledonia
Tonga
Australia
Norfolk Island
Lord Howe Island
New Zealand

Hawaiian Islands

Line Islands

s

Marquesas

Cook Islands

Tuamotus

Society Islands

Raivavae

Rapa Iti

Gambier Islands

Pitcairn Islands

Rapa Nui

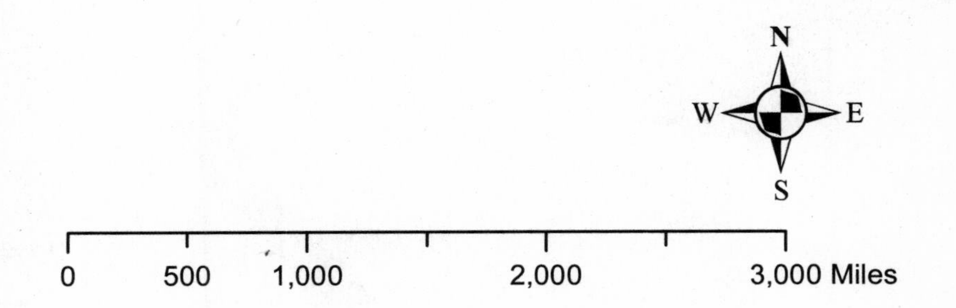

THE STATUES THAT WALKED

UNRAVELING THE MYSTERY OF EASTER ISLAND

Terry Hunt and Carl Lipo

FREE PRESS
New York London Toronto Sydney

fP

FREE PRESS
A Division of Simon & Schuster, Inc.
1230 Avenue of the Americas
New York, NY 10020

First Free Press hardcover edition June 2011

For information about special discounts for bulk purchases, please contact Simon & Schuster Special Sales at 1-866-506-1949 or business@simonandschuster.com.

The Simon & Schuster Speakers Bureau can bring authors to your live event. For more information or to book an event contact the Simon & Schuster Speakers Bureau at 1-866-248-3049 or visit our website at www.simonspeakers.com.

DESIGNED BY ERICH HOBBING

Manufactured in the United States of America

1 3 5 7 9 10 8 6 4 2

Library of Congress Cataloging-in-Publication Data
Hunt, Terry L.
The statues that walked: unraveling the mystery of Easter Island / Terry Hunt and Carl Lipo.
p. cm.
Includes bibliographical references and index.
1. Easter Island—Antiquities. 2. Sculpture, Prehistoric—Easter Island.
3. Prehistoric peoples—Easter Island. 4. Polynesians—Easter Island—Antiquities.
I. Lipo, Carl P. II. Title.
F3169.H86 2011
996.1'8—dc22 2011001627

ISBN 978-1-4391-5031-3
ISBN 978-1-4391-5434-2 (ebook)

To Professor Robert C. Dunnell

(December 4, 1942–December 13, 2010),

mentor and friend, whose contributions to archaeology

and to our thinking made this book possible.

Contents

1. A Most Mysterious Island 1
2. Millions of Palms 19
3. Resilience 33
4. The Ancient Paths of Stone Giants 55
5. The Statues That Walked 73
6. A Peaceable Island 93
7. *Ahu* and Houses 109
8. The Benefits of Making *Moai* 131
9. The Collapse 147
10. Conclusion 177

Appendix 1: Environmental Constraints 181

Appendix 2: Lithic Mulching and *Manavai* 191

Notes 199

Bibliography 211

Acknowledgments 227

Index 229

THE STATUES THAT WALKED

CHAPTER 1

A Most Mysterious Island

The old net is laid aside; a new net goes a-fishing.
—Maori proverb

Mention Easter Island to just about anyone and "mystery" immediately comes to mind. *The Mystery of Easter Island* is the title of untold books and modern film documentaries. The mystery surrounds how so few people on a remote, treeless, and impoverished island could have made and transported hundreds of the eerie, gargantuan statues—called *moai*—for which the island is so famous. The awe-inspiring, multi-ton stone statues, some standing nearly forty feet high and weighing more than seventy-five tons, were carved out of the island's quarry of compacted volcanic ash and then somehow transported several miles over the island's rugged terrain. Not all of them survived the journey. Many lie scattered across the island, some broken, never to take their intended places on platforms along the shoreline or elsewhere throughout the island. To see these statues, many of them situated upon equally impressive platforms called *ahu,* is to sense a hidden drama of compelling human proportions calling out for explanation. Facing inward, rather than out to sea, they seem to be gazing back in a vain search for the noble society that created them.

As we were archaeologists who had studied other parts of Polynesia, when we began our work on the island the statues were somewhat familiar. Similar religious statuary are found elsewhere in Polynesia. And on other islands, statues were also moved significant distances. The *moai,* like the elaborate carved wooden images from the Hawaiian Islands, or the stone tiki of the Marquesas, while much bigger, represented the same deified ancestors so important in Polynesian religion and cosmology. That the *moai* were religious images explains why the vast majority face inland, watching over their descendants day after day. With their backs to the sea, the *moai* had not been carved as sentries, warding off potential intruders, as with the Colossus of Rhodes.

Had the islanders carved and transported just one or two of these statues, the accomplishment would have been noteworthy, but not surprising. But our count for Rapa Nui suggests that the islanders carved something well over 950 statues, and of those, more than 500 were transported considerable distances, appearing in every corner of the island. Nowhere else in Polynesia is such a creative and monumental legacy found. Why did it emerge only on this tiny island, whose population should have, by all accounts, been focused solely on where to find the next meal?

Since Easter Sunday 1722, when the first European accidentally sighted this isolated speck in the vast South Pacific, Easter Island has presented a seemingly intractable dilemma for explorers, scientists, and curiosity-driven tourists. By comparison to the cultural and physical richness of such storied Polynesian islands as those of the Tahiti and Hawaii archipelagos, Easter Island seems a poor setting—almost mocking—for one of the great achievements of early Polynesian history. The island itself, which today Polynesians call Rapa Nui (the people who live there are called the Rapanui), is almost a moonscape in appearance, little more than a barren lump of lava-covered terrain. Lacking the deep valleys, steep mountains, lush streams, and beautiful waterfalls typical of many of the volcanic islands of Polynesia, Rapa Nui is characterized by a modest landscape of rolling hills. The island was born less than a million years ago when the coalescing eruptions of three seafloor volcanoes reached the surface. One searches in vain here for a refreshing stream, let alone a flowing river. Most of the water is found in lakes formed in the three volcanic cones, though some also trickles out of a number of small springs.

Nor does fruit fall from the trees here, as it does on so many other Polynesian islands. From the mid-eighteenth century onward, seamen, explorers, Christian missionaries, and other visitors remarked consistently on the pitiable and "wretched" lives of the island's native inhabitants. Swedish botanist Carl Skottsberg, who compiled the first natural history of Rapa Nui, wrote that "there is in the Pacific Ocean no island of the size, geology and altitude of Easter Island with such an extremely poor flora and with a subtropical climate favorable for plant growth, but nor is there an island as isolated as this, and the conclusion will be that poverty is the result of isolation."[1]

Those who settled Rapa Nui had accomplished a remarkable feat of seamanship, perhaps the most daunting of the whole colonization of the Polynesian islands, only to have arrived at a desperately inhospitable new home.

The story of the Polynesian migration is staggering in its sweep. Seafaring colonists known by their distinctive pottery called Lapita, who had set out from the shores of the western Pacific, reached the islands of Tonga and Samoa by 800 BC. It must have seemed to be the edge of the world. Verdant Samoa is today considered the heart of Polynesia, but at that time, there on the eastern frontier of their rapid dispersal to hundreds of islands, Lapita stopped dead in its tracks. Maybe Samoa was just too luxurious for them to leave. We don't currently know why they stopped, but we do know that no islanders ventured farther into the Pacific for nearly two thousand more years.

It was in the Polynesian homelands of Tonga and Samoa that the earliest forms of Polynesian monumental architecture emerged, by about AD 1000. When the islanders began migrating again, sometime around AD 1100, they brought their ritual architecture with them, including religious courtyards made of stone and upright stones, conceived of as backrests for the gods.

In some places these "backrests" were transformed into elaborate carved human figures, like those found in the Marquesas, Hawaii, the Australs, and, ultimately, Rapa Nui.

Those migrating across the eastern Pacific first reached the spectacular islands of the archipelago of Tahiti. Voyaging in large double-hulled canoes soon after AD 1200, in less than a century, the islanders had discovered just about every island in the eastern Pacific, including the far-flung Cooks, Tuamotu atolls, Marquesas, Hawaii, Australs, Gambiers, Rapa Nui, New Zealand, and even the frigid islands of the sub-Antarctic.[2] They also reached South America, where they fetched the sweet potato and perhaps introduced the humble chicken.[3] Their colonization over this vast region was remarkably fast; they had traversed thousands of miles of turbulent seas, and had done so against prevailing winds and currents.

Discovering Rapa Nui, the most remote of these outposts, was particularly improbable. The territory over which Polynesia spreads is truly vast: about equivalent to the size of the entire North Atlantic Ocean. Roughly 99.5 percent of Polynesia is ocean, and 92 percent of the tiny fraction of land is New Zealand's whopping 112,355 square miles. Beyond the central archipelagos of the Societies, Tuamotus, and Marquesas lies a wide-open expanse along Polynesia's southeastern edge, where the minuscule islands of Rapa Iti, Pitcairn, Henderson, and finally Rapa Nui are found.

Ordinary maps can't convey Rapa Nui's true remoteness. One of the old names recorded for the island, Te Pito o te Henua, translates as the "navel of the world," or perhaps more aptly, the *end of the world*. The island's geographic isolation is magnified many times over by its extreme windward position. Sailing to Rapa Nui from central Polynesia, as the islanders likely did, meant pushing directly into the prevailing east-southeasterly trade winds and correspondingly strong currents of the South Pacific. Doing so would have required tacking, which would have made the journey approximately four times farther than the straight-line distance. If, for example, the islanders had left from the island of Raro-tonga, in the Southern Cook Islands, the tacking distance to Rapa Nui would have been a staggering 12,500 miles.

The trick would be to find enough days of consistent westerly winds. They may well have been aided by El Niño, which reduces the average strength of the east-southeasterly trade winds in the area, bringing westerly wind reversals. Paleoclimatic studies show that about the time the islanders probably set out, El Niño appeared on average about once every four years, so the Rapa Nui settlers may have ridden one of the gusts of regular westerlies that would have been generated.

To have spotted the tiny island was nonetheless quite a long shot. Rapa Nui is tiny, one of the smallest inhabited islands in the Pacific. The total island area measures about sixty-three square miles. The longest east-west axis is just over fourteen miles; the maximum width north-south is less than eight miles. It is possible to walk around the entire island in a day, albeit a long one. This is surely why, as the evidence convincingly shows, the island was colonized only once. Probably traveling in two, or even more, large double-hulled canoes, some thirty to fifty, or perhaps as many as about one hundred men, women, and children embarked on the voyage.

The oral tradition of the island credits the discovery to a chief named Hotu Matu'a. The first signs of land probably came not with actual sighting of the island, but with seabirds returning to their nests flying off to the east at dusk. A lone palm nut or a tree branch floating in the water might have alerted experienced navigators that land was somewhere nearby. They had defied great odds, but their struggle had only begun. The voyagers would have brought with them the critical plants of Polynesian life, including taro, breadfruit, coconut, yams, bananas, sugarcane, turmeric, and kava as well as chickens and small Polynesian rats (*Rattus exulans*), the latter either invited passengers—it is speculated that they were eaten—or as tenacious stowaways drawn to the provisions. Pigs and dogs may also have been on board, although archaeology reveals they didn't make it to the island. The travelers were to face a challenge, though, in bringing their traditional foods to the island. With such scant water for irrigation, which islanders elsewhere in the Pacific used to cultivate taro, the mainstay of their diets, some of the other crops the settlers brought with them could not be grown on the island. Establishing the way of life the islanders were accustomed to would have been a great challenge. And yet this tiny, relatively impoverished island was to become host to the astonishing population of monoliths so admired still today. Here on Rapa Nui, more than a thousand miles from another Polynesian island, more than two thousand miles from the coast of Chile, apparently without influence from any other culture, a prehistoric society emerged that produced some of the most compelling monuments and feats of engineering in all of Polynesia, and perhaps the world. How could that be?

This has been the question sailors, Christian missionaries, self-styled adventurers, scholars, and a slew of other investigators have been asking since the mid-eighteenth century. Over time a consensus developed around the idea that something dramatic had occurred in the past, long before the island was discovered by Europeans, that would account for the miserable state the society was thought to have been in at the time of European contact. A society that had created such monumental statues, the argument went, must surely once have been more noble. But what was that event? When did it happen? And why? With what seemed increasingly compelling evidence, a theory developed that pointed to horrible conflict and twisted priorities within ancient Rapanui society. Modern writers refer to ecological suicide or "ecocide" in their speculations of what unfolded.

The first Europeans arrived on Easter Sunday 1722, when Dutch explorer Jacob Roggeveen sighted the island. The Dutch encountered a treeless island covered with hundreds of giant statues and a population estimated to be about three thousand, described as healthy. The next European visitors arrived forty-eight years later, in 1770, in the form of a Spanish expedition under the order of the viceroy of Peru, which provided little detail about the state of the island other than that there was very little wood. Then, in 1774, the English arrived under Captain James Cook, probably the best-known explorer to have sailed the Pacific. His fame comes in part from the great details of his observations and those of his crew.

The British mission in the Pacific was colonial, and thus economic. Like the other European nations, Britain had aspirations of finding the great southern continent, and as things unfolded, a northwest passage as well. In Cook's first voyage, on the *Endeavour,* he charted much of the Australian and New Zealand coastlines and collected a wealth of information about these southern lands. In July 1772 he set sail from Plymouth, England, for his second expedition, with two ships, the *Resolution* and the *Adventure.* This would be a final search for the elusive southern continent, Terra Australis, and indeed, this voyage was to furnish once and for all proof that it did not exist. On board were German-born naturalist Johann Reinhold Forster and his nineteen-year-old son, Georg, as well as artist and engraver William Hodges.

Searching the frigid waters of the far southern Pacific for about two months, Cook crossed the Antarctic Circle, and as he expected, no great continent awaited their discovery. By March of 1774, the crew, badly in need of provisions, headed north to Easter Island. His men were exhausted and suffering the debilitating effects of scurvy. Cook knew of the voyage of English buccaneer Edward Davis in the 1680s with mention of an island in the vicinity, he had read Carl Behrens's report of the Dutch visit in 1722, and he had learned of the Spanish expedition. On March 11, he and his crew finally sighted Rapa Nui.

Cook and his men described the island as barren, lacking wood and fresh water, and noted, "Nature has been exceedingly sparing of her favours to this spot."[4] On this visit, the expedition's naturalist, Forster, recorded that

> the most diligent enquiries on our part have not been sufficient to throw a clear light on the surprising objects which struck our eyes on this island. We may however attempt to account for those gigantic monuments, of which great numbers exist in every part;

> for as they are so disproportionate to the present strength of the nation, it is most reasonable to look upon them as the remains of better times. The nicest calculations . . . never brought the number of inhabitants in this island beyond 700, who, destitute of tools, of shelter, and clothing, are obliged to spend all their time in providing food to support their precarious existence. . . . Accordingly we did not see a single instrument among them on all our excursions, which could have been of the least use in masonry or sculpture: We neither met with any quarries, where they had recently dug the materials, nor with unfinished statues which we might have considered as the work of the present race. It is therefore probable that these people were formerly more numerous, more opulent and happy, when they could spare sufficient time to flatter the vanity of their princes. . . . It is not in our power to determine by what various accidents a nation so flourishing, could be reduced in and degraded to its present indigence.

This concept of flourishing times on the island followed by indigence took a new twist with French explorer La Pérouse. Visiting for a just single day in April 1786, he speculated that at some very distant time Easter's inhabitants unwisely cut down all of the island's trees. La Pérouse observed that the loss of the forest

> exposed their soil to the burning ardor of the sun, and has deprived them of ravines, brooks, and springs. They were ignorant that in these small islands, in the midst of an immense ocean, the coolness of the earth covered with trees can alone detain and condense the clouds, and by that means keep up an almost continual rain upon the mountains, which descends in springs and brooks to the different quarters. The islands[,] which are deprived of this advantage, are reduced to the most dreadful aridity, which, gradually destroying the plants and scrubs, renders them almost uninhabitable. Mr. de Langle [naval commander, explorer, and second in command of the La Pérouse expedition] as well as myself had no doubt that these people were indebted to the imprudence of their ancestors for their present unfortunate situation.[5]

In this manner, the notion of the imprudence of the islanders' ancestors entered the Western discourse of Rapa Nui's sorry fate, a theme that would be revived in the twentieth century with the beginnings of scientific expeditions to the island.

Best known for his *Kon-Tiki* adventures, Norwegian explorer Thor Heyerdahl brought contemporary archaeological research to Rapa Nui. Born in 1914 in Larvik, Norway, Heyerdahl began his studies as a zoologist and traveled to the Marquesas as part of a school project to learn how animals arrived on these remote Pacific islands. This interest in colonization led him to an interest in human migrations, specifically those of the Polynesians and, ultimately, the Rapanui. Heyerdahl believed that ancient Americans first populated the Pacific islands, and he set out to prove his theory in adventurous fashion.

In 1947 Heyerdahl led the *Kon-Tiki* expedition, a daring experiment, drifting on a balsa wood raft from South America into the Pacific, as some Spanish accounts had asserted the Inca had done. The expedition succeeded in making land, washing ashore on the reef off the coast of the island of Raroia in the Tuamotu Archipelago, and generated a storm of media attention. In 1955 Heyerdahl and an international team of scientists began extensive field research on Rapa Nui. William Mulloy, an American archaeologist, was part of the team. Based on his analysis of the evolution in style of the island's monumental architecture and the island's first radiocarbon dates, Mulloy proposed three periods for Rapa Nui prehistory: the Early Period, AD 400–1100; the Middle Period, AD 1100–1680; and the Late Period, AD 1680–1868.[6] Heyerdahl and his colleagues regarded the Late Period as a time of collapse for the Rapanui civilization; the Early and Middle phases reflected two distinct traditions in an enduring episode of cultural splendor expressed in monuments and colossal statues.

Heyerdahl drew upon the island's oral history for his conjectures, including accounts collected in the late nineteenth and early twentieth centuries.[7] Much of this material was collected by Katherine Routledge, who spent seventeen eventful months on Rapa Nui between 1914 and 1915. Routledge focused much of her research on collecting oral history and observing the islanders' way of life. When writing of her interviews of elder native inhabitants on the island, Routledge remarked with some frustration that it is "even more difficult to collect facts from brains than out of stones."[8] Many of the stories they told have nonetheless been treated as factual by subsequent researchers.

In the oral traditions recorded by Routledge, the island was subject to intense competition between two groups, the so-called "Long Ears" and "Short Ears." In his interpretation of this legend, Heyerdahl argued that the initial colonists of Rapa Nui arrived from South America early in the first millennium AD.[9] At some later, uncertain point in time a second wave of migrants arrived from Polynesia. For Heyerdahl, the South Americans represented a superior "lighter-skinned" population who maintained a tradition of artificially distending their earlobes and thus were known as the Long Ears, while the Polynesians, who did not share this tradition, were the Short Ears. According to Heyerdahl, the early South American settlers also brought with them a specific cultural tradition of highly specialized masonry art. Heyerdahl states in his book *The Art of Easter Island,* "The remarkable expertness of the first settlers suggests a long tradition in stone-shaping technique . . . [and] it is logical to assume that the Early Period settlers brought the art of stone sculpture with them."[10] It was these South American settlers, the Long Ears, Heyerdahl argued, who were responsible for the first *ahu* and *moai* construction.

These opposing tribes, Heyerdahl further conjectured, had clashed in an epic battle around AD 1680, which has come to be known as the "AD 1680 Event," killing all but one individual of the Long Ear tribe. Heyerdahl wrote: "The collapse of the totalitarian hierarchy must have taken place before the arrival of the first Europeans. No supreme sovereign ceremonially received Roggeveen or the Spaniards when they landed. The fact that the Dutch found that peace and apparent equality had been reestablished appears to indicate that the collapse of the organized monarchy must have taken place well before 1722."[11]

Heyerdahl's putative AD 1680 Event still represents a pivotal point in the conventional narrative for Rapa Nui. The great battle of AD 1680 is said to mark the tipping point of environmental and demographic collapse.[12] Heyerdahl's seemingly precise date of AD 1680 comes from a single radiocarbon date that roughly corresponds to an account offered by Father Sebastian Englert, a Catholic priest and researcher who lived on the island from 1935 to 1969. Englert estimated AD 1680 from genealogical records in concert with an array of suppositions. Since the AD 1680 Event derives from little more than interpretations from legends, most contemporary researchers do not believe the story has much in the way of veracity. Nonetheless, incarnations live on in support of the ecocide thesis.

Today the best known is Jared Diamond's popular account in his book *Collapse*. Diamond asserts that Rapa Nui was first settled by Polynesians, around AD 900, or perhaps a bit earlier, and taking account of the most modern analysis of pollen grains preserved in the island's lakes, it was covered in luxuriant forest, a virtual Garden of Eden. Giant palms (*Jubaea chilensis* or a close relative) dominated the forest which, along with the other trees that once thrived on the island, are now extinct.

A complex, hierarchical chiefdom developed on the island with a religion focused on ancestor worship. As things went from bad to worse, religious and political pressure, fueled by greed and shortsightedness, led to more and more statues being carved and transported. The same fervor led to larger and larger statues. Thousands of the giant palm trees were cut down to construct sleds or other contraptions essential to move ever more statues. Thousands more were cut and burned in order to make more room for more and more agricultural fields to feed the hundreds of statue workers. A centralized political authority held a ruthless stranglehold on the island's people and was in charge of the work of the statue cult. Diamond believes the population grew out of control, and by about AD 1600 had reached a staggering 10,000, 15,000, or maybe even 20,000 or 30,000, greatly overtaxing the island's resources.

When the island's resources had become so diminished that

food had become scarce, statue making stopped abruptly—dead in its tracks. Statues were left unfinished in the quarry and many were abandoned along the transport roads en route to display, vivid snapshots of the moment the glory ended. Crisis ensued. Obsidian spear points were manufactured in huge numbers to arm islanders in the fierce battles that broke out in a vicious cycle of violence. Chickens were kept in fortified compounds to stop thieves. Because the forest had been decimated, no canoes could be made for obtaining supplies from other islands. Apart from fish, the only significant source of protein was human flesh, and cannibalism became rampant, further fueling the cycle of warfare.

A great civil war ensued, with thousands killed. To insult their enemies, and perhaps the ancestral gods who had forsaken them, the warring parties toppled the giant statues of their opponents in fits of rage. The real end came when someone cut down the last tree. They knew it was the last tree, and they cut it down anyway. Finally, there was no place to run. The islanders had sentenced themselves to a prison of their own making.

Diamond sees Rapa Nui's demise as "the clearest example of a society that destroyed itself by overexploiting its own resources."[13] Deforestation coupled with environmental fragility and human recklessness brought cultural collapse, and this horrific history is held up as a cautionary tale of environmental ruin that serves as a parable for our own time.

When we began our work on the island, we fully expected our findings to corroborate this account. After all, Rapa Nui is one of the most intensively studied specks of land anywhere in the world. We thought that doing additional field research on the island might add a few details to the already known prehistory, but not much more.

Our first years of fieldwork on Rapa Nui were from 2001 to 2003, when we focused on a large-scale survey of archaeological stone structures and artifacts found on the surface. Our mission was simply to document these remains—many in danger of destruction from land use changes, vehicles, and tourists—while also training our students how to map, photograph, and write detailed descriptions of findings. It wasn't until our fourth field season, in 2004, when we conducted excavations on Anakena Beach, that our understanding of the island's archaeology and prehistory began to change.

Anakena Beach is probably Rapa Nui's most photographed location, a beautiful strip of white sand along a crescent-shaped bay of turquoise water lined with rugged black volcanic rocks. Anakena is splendid by any standard, but the majestic stone platforms known as Ahu Ature Huke and Ahu Nau Nau, both now restored and crowned with well-carved and almost perfectly preserved *moai,* make it an awe-inspiring setting like nowhere else on earth.

In the oral tradition, Anakena was said to be the settlers' landing site, and according to legend, it became the royal residence of Hotu Matu'a, which meant that the beach area would likely be full of records of ancient activity. We expected to find plentiful artifacts, such as fragments of fishhooks, obsidian flakes, and shaped tools, as well as stone adzes and fragments of stone that chipped off during their use. We also anticipated that the calcareous beach sands provided the perfect conditions for the preservation of bone. One of our students doing her doctoral work, Kelley Esh, hoped to study the record of consumption of fish, seabirds, chickens, and perhaps also rats, evaluating how people had changed their diet as resources diminished.

As we expected, the dune deposits were beautifully stratified, with layer after layer of sand periodically interrupted by volcanic mud that had eroded off the slopes above. The stratification formed by consecutive, undisturbed layers of sand and earth meant that we could be confident that the bits and pieces we found in a layer had been deposited together.

Digging in sand dunes is a challenge, because nothing much holds the uppermost sand of a dune in place. The deepest layers in our 2004 excavations were more than ten feet deep, and tourists standing near the edge of our pits to peer down on what we were doing invoked some genuine fear in us. The other excitement would come when our Rapanui friends would gallop their

Figure 1.1. The beach setting at Anakena and the *ahu* known as Ahu Nau Nau.

horses over the beach and up dunes. Down in the pits we could hear them coming as sand grains began to fall all around us. We had to continually build wooden frame reinforcements around the sides of our pit.

The work paid off, though, as we found abundant artifacts and bones, an archaeologist's dream. In 2005 we returned to Anakena and renewed our work, this time in an area that had the same layers, but where they could be found much closer to the surface. Again the ancient sandy layers offered up thousands of small bones—from birds, fish, rats, and chickens, and fragments from sea mammals such as dolphins—as well as obsidian flakes and some ground and polished stone adze fragments.

In both the 2004 and 2005 excavations we eventually reached a dense, compact clay layer, in which we could see charcoal bits, obsidian flake artifacts, and bones, mostly from rats and fish, clear evidence of human habitation, as none of these things would be embedded in this ancient surface naturally. But as we excavated below that level, we found none of that: no charcoal, no obsidian flakes, and no bones. We did see small tubular casts in the clay that we quickly recognized as what are called root molds, empty voids preserved where the roots of ancient plants once grew. These root molds provided direct evidence of the storied giant palms. We had uncovered the soil in which the palms grew when people first set foot on the island.

On a sunny and windy day, as our work was winding down, we cleaned the sand and any debris away from the clay layer in preparation for carefully taking samples for radiocarbon dating. Painstakingly we collected small bits of charcoal from the ancient soil, expecting that these samples would date back to about AD 700 to 800, based on the chronology we believed was well established for the island.

Some archaeologists still accepted Heyerdahl's earlier date of AD 400 for colonization, but we, like a growing number of others, saw serious problems with Heyerdahl's date. No one had been able to replicate it from charcoal samples taken from the same layer Heyerdahl had taken his samples from. And worse, Heyerdahl's AD 400 date was measured from a sample of unidentified wood charcoal, which probably led to what radiocarbon experts refer to as the "old wood" problem: because the rings of a tree's interior are older than those of the exterior, the parts of the wood dated can produce different ages. This is why we now date wood charcoal from short-lived trees or short-lived parts such as twigs or seeds. The emerging consensus was for a colonization date at least a few hundred years later than Heyerdahl had thought.

Months passed and finally the radiocarbon report arrived. The results were consistent, the statistical errors were small, and the laboratory reported that everything had proceeded normally. But there was a problem. The oldest layers at the base of Anakena were only around eight hundred years old. Eight hundred years? That would mean that the colonizers had arrived much later than even the newly accepted later date, of AD 700 to 800, thus in the neighborhood of only about AD 1200. Had something gone wrong? We were disappointed. Feeling puzzled, we set the report aside and went on to other tasks. But before long we began to rethink what the evidence might be telling us. Could this "late" date be correct? Was Rapa Nui settled centuries later than everyone had assumed? We e-mailed Atholl Anderson, a leading expert on the cultures of the Pacific islands, asking him what he thought, and he responded that a date of about AD 1200 would actually fit the picture emerging for the wider settlement of East Polynesia much better than the consensus older dates. He also added words of sage advice: "trust the hard evidence more than your preconceptions."

But if our radiocarbon dates were right, then what about the older ones that existed for Rapa Nui? Were there problems that had been overlooked in obtaining these dates? Were some of those charcoal bits perhaps also actually portions of ancient trees that began their life hundreds of years before humans arrived? We decided to conduct a critical evaluation of all of the more than 120 radiocarbon dates published for Rapa Nui,[14] excluding dates measured from any problem materials such as unidentified wood charcoal. We also excluded marine samples such as shell, fishbone, coral, and even seabird bones, because sea creatures often feed in the areas of oceans where upwellings from the deep open create high nutrient levels, which means that they absorb ancient dissolved carbon dioxide that has been released into those upwellings. Consuming this old carbon makes marine samples appear older, sometimes by centuries.

Our analysis showed that all of the reliable dates—those from short-lived materials—fit with our new evidence. Several pieces now fell into place supporting the AD 1200 chronology for settlement. Recently acquired dates from new pollen studies on the island had shown the first signs of human presence to be at about the same time,[15] and recent studies also showed sudden environmental changes around AD 1200. New research was also confirming later chronologies across Polynesia, in Tahiti, the Marquesas, New Zealand, and Hawaii.[16] We knew that our results were important, and we immediately wrote a paper introducing them, which within a matter of weeks appeared in the journal *Science*. On Rapa Nui, as elsewhere in Polynesia, shorter chronologies meant rapid disappearance of the native forest occurred soon after colonization. With careful dating, the major changes in island vegetation and the first—independent—signs of human presence were closely contemporaneous. The emerging evidence pointed to significant forest loss on the scale of decades, not centuries. For Rapa Nui and other Pacific islands, this evidence raised questions about how the forest could decline so rapidly. We grew skeptical that the efforts of making and moving the giant statues explained much about forest depletion on Rapa Nui.

If something as fundamental as the dating of first colonization of the island had been wrong, even several hundred years off, what else didn't we know? Critical thinking had led us to doubt a few details, but the shortened chronology now undermined our confidence in just about everything. It felt as though what we knew, or thought we knew, had been pulled out from under us.

We realized that there was a lot of work to do, and that things might just turn out quite differently than we had assumed. As we began to study what was actually known, versus speculation that had been passed off as certainty, we began to suspect that the story needed revision.

We had no question that deforestation had transformed Rapa Nui, but we now wondered how and why this deforestation occurred and whether it had actually caused a population collapse. That was to be our next focus of investigation.

CHAPTER 2

Millions of Palms

> Easter Island has the dubious distinction of being one of the most misinterpreted and misunderstood areas of its size in the world.
>
> —Patrick McCoy, *Easter Island Settlement Patterns in the Late Prehistoric and Protohistoric Periods,* 1976

Seeing Rapa Nui today, it is so barren and exposed it is hard to imagine that the island was once covered with a dense palm forest. Reconstructing the nature of the island's forest and explaining the timing and causes of its dramatic loss would be critical to understanding the whole story of the island, and we set out to determine exactly how the deforestation figured in the island's alleged collapse.

A variety of research now shows that when Polynesians first set foot on Rapa Nui it was indeed covered in a forest of literally millions of giant palms. Similar to their cousins that survive on the Chilean mainland, these stately trees towered to heights well over one hundred feet, making them among the largest palm trees in the world. They grew slowly, perhaps taking as much as a century to reach maturity and bear fruit, and then living for hundreds more years. The first scientific evidence that the island had once supported a forest was produced by Swedish botanist Olaf Selling, who analyzed the microscopic pollen grains preserved in lake bottom sediments in samples taken by members of the Heyerdahl

Expedition. During the same expedition, excavators also mention finding the telltale root molds left deep in the soil by ancient roots that we had encountered.[1] From their later excavations on the island, William Mulloy and Gonzalo Figueroa also noted the presence of "occasional tunnel-like tentacles . . . interpreted as molds left by root activity" and concluded that the island "was once covered with significantly more vegetation."[2]

That the lush forest had vanished was of no doubt, but we had many questions about how quickly that had happened and why. On the subject of deforestation, Jared Diamond had nicely summed up the conventional story this way:

> Eventually Easter's growing population was cutting the forest more rapidly than the forest was regenerating. The people used land for gardens and wood for fuel, canoes, and houses—and of course, for lugging statues. As forest disappeared, the islanders ran out of timber and rope to transport and erect their statues. Life became more uncomfortable—springs and streams dried up, and wood was no longer available for fires. . . .[3]

For his account, Diamond[4] had drawn on the work of archaeologist and popular author Paul Bahn and paleoecologist John Flenley, who had addressed the same issue in their book *Easter Island, Earth Island* by conjecturing that

> the person who felled the last tree could see that it was the last tree. But he (or she) still felled it. This is what is so worrying. Humankind's covetousness is boundless. Its selfishness appears to be genetically inborn. Selfishness leads to survival. Altruism leads to death. The selfish gene wins. But in a limited ecosystem, selfishness leads to increasing population imbalance, population crash, and ultimately extinction.[5]

We had grown skeptical of this interpretation. So we decided to take a close look at all of the evidence about what might have happened to the island's magisterial palms.

The earliest descriptions of the environment of the island turned out to be contradictory. The account written after the first expedition to observe the island, that led by Jacob Roggeveen in 1722, had described the island as appearing, from a distance, as "being of a sandy nature . . . because from its outward appearance it suggested no other idea than that of an extraordinarily sparse and meager vegetation."[6] But the account also went on to say that once they arrived on the island,

> we found it not only not sandy but to the contrary exceedingly fruitful, producing bananas, potatoes, sugar-cane of remarkable thickness, and many other kinds of the fruits of the earth; although destitute of large trees and domestic animals, except poultry. This place, as far as its rich soil and good climate are concerned, is such that it might be made into an earthly Paradise, if it were properly worked and cultivated; which is now only done in so far as the Inhabitants are obliged to for the maintenance of life.[7]

The Dutch crew also mentions seeing "small coconut palms,"[8] houses "covered in palm leaves,"[9] and "in the distance whole tracts of woodland."[10] At this point, it would seem, the island was still fairly rich in vegetation and still boasted some of its lush palm forest. Even 140 years later, in 1868, when British naval surgeon J. Linton Palmer visited the island, he observed "boles [trunks] of large trees, Edwardsia, coco palm, and hibiscus, decaying in some places."[11] He describes the large trees as coconut palms, but we learned that coconut trees were not introduced to the island until the nineteenth century, so Palmer must actually have been describing the last remnants of the *Jubaea* palms. These descriptions, therefore, suggested that at least some of the trees survived past the European discovery of the island, a possibility that some other researchers had mentioned, and therefore that the process of deforestation was perhaps not as relentless as the conventional story depicted.

We then began examining the best modern evidence regarding when deforestation occurred. Perhaps by pinning down a definitive date for that, we would begin to develop a more robust understanding of just what the human impact on the primordial environment had been, and how quickly degradation had happened.

John Flenley, along with researcher Sarah King, had analyzed three sediment core samples from different parts of the island in 1977 and assembled the first detailed evidence for deforestation.[12] The first was from Rano Aroi, a swamp in a crater near the island's summit; the second from Rano Raraku, a crater lake directly adjacent to the main statue quarry; and the last from the large, deep lake in the Rano Kau crater. The pollen evidence confirmed that a palm forest had been replaced by grass, but the evidence for timing of its disappearance, and what role the islanders might have played, was less definitive.

Flenley's analysis of the core from Rano Raraku, the lake adjacent to the main statue quarry, seemed to show a dramatic change below a layer dated to about 500 years ago.[13] Before that, the pollen data indicated the giant palm were abundant trees, but from about 500 years ago, AD 1500, and continuing to the present, the deposit revealed no signs of these native plants. The analysis of another core taken from the lake of Rano Kau crater—reaching thirty feet in depth—corroborated this finding. Flenley concluded that the forest clearance began by at least AD 800, and that the last remnants of the island's forest had been destroyed by AD 1500. However, in recent publications Flenley and his associates reported major vegetation changes as early as around AD 100, and they speculated that this may mark the first signs of human arrival. Based on different studies, the island's chronology for colonization had come to vary by more than a thousand years.[14]

The discrepancies can be explained by a problem with the lake sediment cores. In a more recent analysis of radiocarbon dating of samples from the cores, researchers demonstrated that old and young materials can be mixed together.[15] This kind of mixing would explain the major problems in reconstructing the timing of vegetation change. It turned out that the association of pollen evidence for vegetation change and the samples used for radiocarbon dating proved highly unreliable. Samples from the same layers in the core were yielding results thousands of years apart and sometimes in reverse chronological order. Additional careful work was necessary, and fortunately a good deal of new research has been done in recent years.

Between 1997 and 1999, paleoecologist Daniel Mann and his associates collected a sediment core from the lake at Rano Raraku.[16] Once again the pollen record clearly showed the decline of a palm forest and its replacement by grassland. His team's analysis also showed clear evidence of soil erosion on the slopes beginning after AD 1200. At about the same point in time the evidence also shows a massive influx of charcoal particles. These findings painted a picture of people using fire for the first time, palm trees disappearing from the slopes of the crater, and soil erosion following the loss of vegetation.

Mann and his colleagues then conducted studies of soils from nineteen coastal and interior locations across the island. They found evidence that the island's primeval soils, which were riddled with root molds from the ancient palm forest, were often overlaid by eroded soils with charcoal particles. Their analysis of the eroded soils yielded dates spanning a period from about AD 1280 to about 1650, and they concluded that after AD 1200, the island's original soils were eroded throughout the island, suggesting that forest loss had occurred over that 370-year time frame from AD 1280 to 1650.

The Mann team's results found support in separate work done by French researcher Catherine Orliac and a German team of ecologists headed by Andreas Mieth.[17] Orliac compiled a remarkable sample of thousands of carbonized fragments of wood, seeds, fibers, and roots from three locations across the island, dating between about 600 to 200 years ago. Her results offered a close match, showing that from about AD 1300 to 1650, or a bit later, many forest plants still grew on the island. But by about 1650 most of the island's wood had disappeared. Evidence from the remains of earth ovens also shows that from that time forward, with little wood for fuel, other plants, such as grass, ferns, and sweet potato vines, had been burned. Beginning in 2002, Mieth's team conducted research on the island's Poike Peninsula as well as several other locations around the island, again finding strong evidence that a dense palm forest had covered most of the island, with their dates pointing to deforestation evident at about AD 1280.[18]

All of this field research, from multiple locations around the island, has provided a coherent picture, with the effects of human impact consistently showing up about AD 1280, which was further strong support for our later date of colonization. At the time the studies were done, though, some of the researchers were still assuming that the islanders had arrived much earlier, which left them with a conundrum to solve. Why, if they had arrived in AD 800, had deforestation only begun so much later? Mann and his coauthors[19] suggested that little land was cleared for so many years because the population had remained small. But even as they made this conjecture, they pointed out that it was unlikely to be the full answer, because the island has a high potential for wildfires, and those on their own would likely have devastated more of the forest earlier than their evidence showed. So they postulated another explanation: that the island was occupied only intermittently between AD 300 and 1200.[20]

Andreas Mieth's team offered another explanation.[21] They suggested that for the first several centuries of settlement, the islanders managed a kind of "low impact" sustainable economy, but then at about AD 1250 they envisioned something very dramatic unfolding to radically upset the balance. Their speculations were vague: a "new cultural impetus" initiated the "unstoppable process of degradation."

Of course, with our revised date of colonization, the onset of deforestation at about AD 1200 was no longer puzzling. But some of the research suggested almost complete devastation within just two hundred years, while other studies indicated that deforestation occurred over a longer time frame, which fit well with the observations made by the Dutch in 1722. In any case, one thing was perfectly clear: deforestation had begun less than a century after arrival, so there was no extended period when the islanders had little to no impact on the native forest. These findings, and our later date for Polynesian arrival, might have provided strong new support for the conventional story—the islanders recklessly cut down trees and continued to do so at a rapid rate even after serious depletion would have become clear. But we thought otherwise. Reason for doubt came from our knowledge of some fascinating new research about the deforestation of the Hawaiian Islands. We expected that the fate that had befallen Hawaii had also visited Rapa Nui.

Archaeologist Steve Athens, who directs the International Archaeological Research Institute in Honolulu, and his colleagues have done remarkable field research in many parts of Hawaii, as well as elsewhere in the Pacific, concerning the impact of ancient humans on island environments. We've known Athens for a long time, and we were aware that since the 1980s he had been studying the deforestation of the Hawaiian Islands, where up to 90 percent of the native lowland forests were lost by about AD 1500. The prevailing explanation had been that islanders practiced slash-and-burn agriculture, burning more and more forest to convert it into fields. But Steve had made note of a strange puzzle. There was very little evidence of charcoal in the sedimentary records. If the islanders had done so much burning, the sediments should have been full of charcoal. Looking back, it is hard to comprehend how others of us in the field could have failed to make note of this glaring contradiction, but we were so sure we had the right answer in slash-and-burn that we simply didn't dig deeper. Fortunately, Steve did, and what he discovered was striking.

In 2001, Terry invited Steve to give a guest lecture to his class about recent archaeological work on the Ewa Plain—a large, flat area formed from ancient coral that is on the southwestern corner of the island of Oahu. Today the Ewa Plain comprises the outer suburbs of Honolulu, and much of it is still undergoing rapid construction of houses, roads, and shopping centers. The flurry of development has resulted in much archaeology, as federal and state laws require some degree of investigation before construction begins on previously unmodified lots. In the lecture that day, Terry learned that Steve had come up with a compelling new answer to the mystery of the devastation of the lowland forests that had once covered the Ewa Plain.

Athens and his team had analyzed archaeological and paleontological samples from the numerous limestone sinkholes dotted around the Ewa Plain, which had accumulated a great deal of bones, sediment, and other ancient environmental remains over thousands of years. They had also performed surveys and excavations of a number of other prehistoric sites, and taken a sediment core from a small brackish lake known as Ordy Pond, located along the southern coast of Oahu. Over time, dirt, charcoal, and plant fragments such as leaves, twigs, and pollen have fallen and washed into this small pond and have accumulated, forming many discrete layers. Drilling down deep and collecting a core of the accumulated muck, researchers can analyze pollen grains, charcoal particles, seeds, and so on in order to reconstruct the vegetation, the fire history, and the changes in the vegetation.

Athens's group collected a deep sample from Ordy Pond that consisted of an impressive twenty-six continuous feet of finely laminated sediment that reflects the environment from the present all the way back to AD 450, a time known to be well before the Polynesian colonization of the Hawaiian Islands. The frequencies of pollen grains showed a rapid disappearance of the dry, lowland forest, with the disappearance of the dominant *Pritchardia* palm and a woody shrub called *Kanaloa kahoolawensis.* The forest had dramatically declined in less than one hundred years, sometime after AD 1200, just about the time Polynesians first arrived in the islands.

The charcoal in the core collected from Ordy Pond revealed a detailed history of fire, for the local area as well as for the island of Oahu more generally. The earliest signs of fire show up as microscopic charcoal particles blown from other locations on the island at depths in the lake sediment corresponding to sometime after AD 1100–1200 (the dating leaves some wiggle room). This evidence indicated that Polynesians had arrived, but that they had not yet arrived in this arid and thus less desirable part of Oahu. A century or so later, larger charcoal particles, ones that could not be blown from any distance, appear in the Ordy sediments. These larger particles document fires in the local area around the pond. But these local fires began only *after* pollen evidence indicates that the forest of *Pritchardia* palm had all but disappeared. The remarkable conclusion: deforestation preceded fire. Thus the native vegetation had not been destroyed by fire. Like the evidence Athens had seen in so many other places on Oahu, here again was the precipitous decline, indeed the crash, of the forest, but it could not be blamed on fires used in Polynesian slash-and-burn agriculture.[22] Humans were not at the scene of the "crime." The picture was strange, but well documented. Hawaiians had settled the Ewa Plain only after deforestation had occurred. Athens's team found this to be true for many other areas of Hawaii as well, and the data also indicated that it was the loss of forest, not necessarily hunting, that had led to the extinction of many native birds and other species.

What Athens proposed was radical: the Polynesian rat had caused the rapid demise of the forest. Bizarre as that theory sounded at first, his evidence was extraordinary.[23]

The combined evidence for fire history from Ordy Pond, radiocarbon dates of rat bones from the Ewa Plain sinkholes, and dates from ancient settlements showed that rats had arrived on the plain about a century or so before humans had actually settled there, when other parts of the island were occupied, but not yet the plain. The appearance of the rats coincided in time with both the beginning of the forest depletion and the commencement of bird extinctions. And an important clue was the high density of rat bones found in the sinkholes, suggesting that the rat population had grown quite large. This made good sense, as the rats had arrived in the Hawaiian Islands with few, if any, predators. The rats also found little competition for food from native birds, since they are agile climbers and, unlike birds, have teeth. They can penetrate the hard, thick seed cases of native plants, which Athens argued they had done prodigiously. Indeed, seeds are among the main food for the Polynesian rat, though they do also eat birds and their eggs. A great deal of ecological research over time had shown that rats can have profound effects on vegetation, though this information was not widely known among archaeologists.

Ecological studies have also revealed that rats tend to reproduce at phenomenal rates. Where food is plentiful their populations can erupt with astonishingly rapid growth. With this piece of information, the last piece of the puzzle came into place. Athens concluded that the deforestation of the plain was a classic case of ecological devastation by an invasive species. The rats ate the seeds of native plants, including the nuts of the native *Pritchardia* palms, and that in turn effectively stopped the natural regeneration of trees. As fewer and fewer new trees grew, and more and more older trees died off, the forest rapidly thinned. The stunning realization was that rats alone were capable of instigating massive, even rapid deforestation. Neither fire nor felling of trees was necessary, even though Polynesians certainly transformed their islands for agricultural production.

The dramatic story of invasive rats has been documented elsewhere. Those who have studied New Zealand know it well, as much field research on the northern offshore islands has documented the serious impacts of rats on forest vegetation and extinction of native animals. Lord Howe Island is another well-documented, ongoing case of rat impact. A strikingly beautiful island located 370 miles off the east coast of Australia, apparently never colonized by Polynesians (or the Polynesian rat they so often carried with them), it was first discovered in 1788 by the crew of the HMS *Supply*, sailing between Botany Bay in Australia and Norfolk Island north of New Zealand. The island is a World Heritage site in recognition of its unique plant and animal life. Unfortunately, black rats were accidentally first introduced to the island when a steamship ran aground in 1918.

By 1921, soon after their invasion, naturalist Allan McCulloch wrote that "this paradise of birds has become a wilderness, and the quietness of death reigns where all was melody."[24] Indeed, rats are now implicated in the loss of five endemic land bird species, thirteen invertebrates, and significant reductions in native plants. Rats continue to be a serious threat to at least thirteen other bird species, two reptiles, and fifty-one plant species, in twelve native vegetation communities remaining on the island.[25]

The native palms unique to Lord Howe Island, a group known as *Howea,* provide an illustrative case in point. The palms, also known as Kentia, are attractive and popular houseplants. Kentia palm seedlings are raised on the island and regularly exported as an important part of the local economy. A recent field study of two palm species on Lord Howe compared seed losses and seedling growth between areas baited to kill rats and those left without bait.[26] The results revealed the dramatic impacts from rats. In baited areas where rats were poisoned, small juvenile palm seedlings were common, but they were rare in the areas free of bait. For the smaller seed palm species, as much as 100 percent of the seeds were lost to rats, and for the other palm, with larger seeds, 20–54 percent were lost. The larger seeds were more difficult for rats to consume. These researchers concluded that rats could quickly lead to the extinction of these palms by consuming their seeds and seriously depressing or even halting new generations. Significant impacts and the road to extinction from invasive rats couldn't be much clearer.

In efforts to combat the invasion of rats on Lord Howe, nonnative owls were introduced to the island between 1922 and 1930, and though the owls ate many of the rats, as planned, they sadly also preyed upon native birds. Eradication efforts are now under way, with plans to dump forty-two tons of rat poison over the entire island in hopes of saving the remaining wildlife.[27]

Athens's work, combined with this other evidence, suggested a compelling new explanation of what happened to the forests of Rapa Nui. After the Polynesian colonists arrived on the island, in about AD 1200, the rats they had brought with them gorged themselves on the unlimited food supply. Millions of giant palm trees offered a virtually endless buffet of nuts. Laboratory studies of rat reproductive potential under such ideal conditions document that populations can double every forty-seven days.[28] Starting with only a single mating pair, the rats could have attained numbers into the millions in just two years or so.[29] Eventually the rat population would have gone into decline, adjusting their numbers to the amount of available food.

The ecological case for this scenario is so strong that one has to ask how the rats could *not* have brought dramatic devastation on the *Jubaea* palms, and other native plants, as well as preying of course on nesting seabirds, land birds, and their eggs and chicks, in addition to snails and many insects, dramatically and rapidly transforming Rapa Nui's ecology.

We were to learn, in fact, that the idea that rats played a major role in Rapa Nui's deforestation was not entirely new. Anthropologist Grant McCall told us that in 1968 he recalls the resident Catholic priest and researcher on Rapa Nui, Sebastian Englert, telling him about "little coconut seeds" (*Jubaea* palm nut shells) he had found in "rat caves." That may have been the first inkling of the role the rats had played. Botanist John Dransfield, who joined John Flenley in early pollen studies on Rapa Nui, also noted the discovery of nuts of the extinct palm in caves, which had been gnawed by rodents, and they argued that this "could have helped to make the species extinct."[30] And John Flenley and his associates had also hypothesized that "the effects of introduced rodents on the biota of oceanic islands are known frequently to have been disastrous . . . and it seems that Easter Island may have been no exception. Whether the extinction of the palm owes more to the prevention of regeneration by rodents, or to the eating of the fruits by man, or to the felling of the mature trees, remains an open question."[31] That question had simply not been followed up on.

Some may wonder why rats would have such devastating effects on Rapa Nui, as they don't always cause such forest destruction. When visiting the native palm forests of mainland Chile, for example, at the natural preserve at La Campana National Park, one is struck by all the palm nuts that have been consumed by native rats, with telltale gnawing marks, and yet here the palms and rats have coexisted for tens or hundreds of thousands of years. The answer is that the impact that rats have depends on a number of factors. First, continental ecosystems are vastly more complex than those on islands generally. In mainland Chile, there are a number of predators for the native rats—such as reptiles and birds of prey—that keep the rat population in check, whereas on islands rats often lack predators. Also, in oceanic island ecosystems, birds have been the main players in plant-animal coevolution; they are the main predators, and rats have been, in most cases, the first creatures with teeth to arrive on remote islands. Unlike birds consuming seeds with their beaks, consuming with teeth destroys many of them.

Of course many islands have native forest that has persisted despite invasion by the Polynesian rat. But as biologists tell us, each island is unique in its history, biodiversity, and biogeography. To argue a false and simple cause-effect that "rats mean deforestation" would assume that diverse islands share the same history and ecology. Jared Diamond said it well: "rats have caused catastrophic extinction waves on some islands, a few extinctions on others, and no visible effect on still others."[32]

Having learned about all of these findings, we set out to determine what the evidence would tell us more specifically about rat devastation on Rapa Nui. We weren't going to assume that rats accounted entirely for Rapa Nui's deforestation, even though left alone they probably could have. The remaining question was what were the relative impacts of rats, fires, and the felling of trees by the colonizers on Rapa Nui's deforestation.

We know from early historic visitors that the deforestation was complete or nearly so by about AD 1722. A dense forest of palm trees and more than twenty other woody tree and shrub species had mostly disappeared. At least six land birds, several seabirds, and an unknown number of other native species were lost to extinction. Many of these losses occurred before Europeans reached the island. Other extinctions probably came after livestock were introduced to the island by Europeans, causing environmental abuse. By the late nineteenth century thousands of introduced sheep, cattle, and horses were grazing freely over the island and would surely have delivered a final blow to native plants or animals that may have survived.

For Rapa Nui we have seen that deforestation began shortly after Polynesian arrival, but loss of the native forest took four hundred years, perhaps longer, with remnants of the vegetation lasting long enough to be witnessed by Europeans. Over this same period, Polynesian numbers increased to a maximum population of three thousand or so by about AD 1350, even as the forest decreased. While colonists on Rapa Nui would face other problems, deforestation did not, therefore, spell disaster. Nor was the story one of people recklessly cutting down the island's last tree.

The best evidence suggested to us that one of the primary rationales for the ecocide scenario of the fate of the island was badly flawed. A great deal about the nature of the native culture had been inferred from the notion that the islanders had been reckless in their management of the forest. If they had, in fact, had so little to do with its depletion, we believed it was now imperative to fundamentally rethink the story of what sort of environmental stewards they were. We also wondered how much of the rest of the story might turn out to be different.

CHAPTER 3

Resilience

> We have seen that neither the position nor the fertility of Easter Island can account for its extraordinary outburst of memorial art. Yet difficulties to be met with often develop exceptional talent, and a hard environment has often bred a people of fine courage and capacity.
>
> —John Macmillan Brown,
> *The Riddle of the Pacific,* 1924

In 1774, Captain Cook was chagrined that Rapa Nui could provide so few provisions for resupplying his ships. He remarked, "there can be few places which afford less convenience for shipping than it does. There is no safe anchorage; no wood for fuel; nor any fresh water worth taking on board. Every thing must be raised by dint of labor, it cannot be supposed the inhabitants plant much more than is sufficient for them; and as they are but few in number, they cannot have much to spare to supply the wants of visitant strangers."[1] Cook limited his visit to the island to just three days.

All the early European visitors made such observations of the resources on the island, as all ports of call in the Pacific were important for resupply. Recall that Dutch captain Jacob Roggeveen noted during his 1722 visit that the land produced "bananas, potatoes, sugar-cane of remarkable thickness, and many other kinds of the fruits of the earth; although destitute of large trees and domestic animals, except poultry." The chief pilot of the Spanish fleet that arrived in 1770 reported that his men "saw no kind of wild nor domestic animal, excepting hens and some rats. The fields are uncultivated save some small plots of ground, in which they sow beds of yucca, yams, sweet potatoes, and several plantations of plantains and sugar-cane: but all very tasteless, as if from want of cultivation." Johann Forster, the naturalist on Cook's visit to the island in 1774, reported, "the whole number of plants growing upon [this island] does not exceed twenty species."[2]

It is clear that at the time of European contact, the array of food available on the island was quite limited, and the archaeological evidence shows that this was true well before the Europeans arrived. Botanists Catherine and Michel Orliac have conducted field and laboratory research designed over the last decade to determine the composition of plant species on the island in prehistory.[3] Through this laborious work, the Orliacs identified twenty taxa of woody plant species, including the Easter Island palm (*Jubaea chilensis*), bushes such as *Sophora toromiro,* and other types of shrubs and small trees. Their samples also contained food remains, and a collaborator of theirs, Erik Pearthree, identified a range of foods consistent with those noted by the early European visitors: sugarcane, taro, *ti* leaf (a leafy plant that is grown for its large, waxy leaves), sweet potato, and yams.

Excavations at Anakena Beach, the sands of which are particularly conducive to preserving bone, have revealed that the islanders also ate a mix of fish, birds, and animals that included dolphins, seals, sea turtles, fish, seabirds, land birds, chicken, and rats. We would expect to find this diet on the Pacific islands, but two key foods that we would also normally see are missing: the island apparently lacked both pigs and dogs. The Polynesians generally brought dogs and pigs, along with chicken and rats, when they set out to colonize other islands, as they were important sources of protein. We also found that most of the fish in our excavations were limited to those that inhabit the near shore waters; and these were limited since Rapa Nui lacked coral reefs so productive in other parts of Polynesia. So the diet on Rapa Nui was significantly less rich than on most of the other Pacific islands, and that was true right from the start of colonization.

The most abundant animal bones found in the excavations are those of the Polynesian rat. They composed roughly 60 percent of all the faunal remains that we collected in our own excavations. This lends credence to the belief that islanders brought rats with them as food. We know that elsewhere in Polynesia, rats were eaten, and there is also a reference to the Rapanui doing so in an account of a visit to the island by Georg Forster, the naturalist on Cook's expedition. He writes, "They also have rats, which, it seems, they eat; for I saw a man with some dead ones in his hand; and he seemed unwilling to part with them; giving me to understand they were for food."[4]

So we have relatively detailed accounting of the plant and animal resources available to the prehistoric Rapanui population and the composition of their diet. And given this fairly impoverished diet, what has long been puzzling was that the archaeological record seemed to contain no obvious evidence of large-scale prehistoric farming. Nowhere on the island can one find the remains of extensive terracing, for example, which might be expected, as we do find them on other islands with similar kinds of environments. On the northern part of the island of Hawaii, large prehistoric field systems are clearly visible,[5] and we know that these fields enabled the prehistoric (that is, AD 1400–1800) population to cultivate sweet potato, yams, taro, bananas, and other nonnative plants, a variety similar to that found on Rapa Nui. These prehistoric Hawaiian farmers also constructed an extensive series of low parallel earthen and stone walls that shielded their crops from the winds that blow vigorously over the slope of the island. Those walls also reduced the loss of water due to evaporation. In fact, based on studies of the effects of windbreaks on evaporation, we can say that they may have resulted in a 20–30 percent reduction.[6]

We wondered why the prehistoric farmers of Rapa Nui hadn't done the same. Given their prowess in transporting multi-ton statues across the island, it would seem that they were plenty

Figure 3.1. View of the Kohala field system showing an expanse of prehistoric agriculture on the northern part of the island of Hawaii.

capable of such engineering. Indeed, the early European explorers were struck by how little effort the islanders seemed to invest in such means of increasing cultivation. Numerous visitors made keen notes on the potential of the island based on the perspective of what Europeans would do. In his notes from his 1722 visit, Roggeveen wrote that "this place, as far as its rich soil and good climate are concerned, is such that it might be made into an earthly Paradise, if it were properly worked and cultivated; which is now only done in so far as the Inhabitants are obliged to for the maintenance of life."[7]

The Rapanui seemed to be underutilizing the island, and this was all the more perplexing given the enormous amount of effort they had apparently put into making their massive statues and stone platforms. This apparent contradiction was behind the conclusion that the islanders were living on the edge of survival in the aftermath of some past calamity. A culture that could produce such monumental works, the argument went, ought to have intensively managed field systems, and should have had concentrations of the population in villages, with land being reserved for cultivation and food production capable of supporting large populations. There seemed to be no other explanation: something very bad must have happened to these people.

It was this perception that led the French explorer La Pérouse—the next European explorer to arrive on the island after Captain Cook—to bring food supplies and new cultigens to the people of Rapa Nui. He set sail from France in 1785 on a round-the-world mission of exploration sponsored by King Louis XVI, with two ships, the *Astrolabe* and the *Boussole.* Arriving on Rapa Nui on April 9, 1786, he spent just a single day on the island, making observations and trading with the islanders, leaving them goats, sheep, pigs, and a wide array of plants, including cabbage, beets, maize, pumpkins, orange trees, lemon trees, and cotton.

But however well intentioned, his gesture was ill conceived. We now know that his scheme was destined for failure. Indeed, the animals were quickly consumed and the plants either failed to grow or quickly dwindled.[8] Were the islanders fools not to have made better use of them? No, the problem was that the island simply wasn't an environment suitable for sustaining the breeding of animals or cultivating such crops.

The landscape of Rapa Nui has little resemblance to the Dutch, English, Spanish, or French countrysides. The island is made from the weathered remains of ancient volcanic eruptions. Despite a somewhat tropical location, rainfall is seasonal but neither abundant nor predictable. While it is possible to attempt most forms of cultivation on the island, it is clear that few of them will be successful over the long run. The environment is so impoverished, in fact, that rather than seeing the islanders as environment destroyers, we would argue that they should be seen as ingenious environmental stewards. They might well have succumbed to the island's impoverishment of resources. That happened on other Pacific islands. Polynesians inhabited Pitcairn Island in prehistoric times. Yet when the mutineers from the *Bounty* arrived on the island in 1790, it was uninhabited. The same was true for scores of other islands across the Pacific, including remote Necker and Nihoa in Hawaii, Howland in the Phoenix Islands, and Washington, Fanning, and Christmas in the Northern Line Islands.[9]

Not all people on all islands managed to sustain their existence as those on Rapa Nui did. But how did they do so? If it wasn't by building terraces and protective walls, what were their methods?

Careful examination of the archaeological record provides the answers: they made good use of two techniques—the building of rock circles, known as *manavai,* and extensive rock-mulch gardening. Though both are well known from the archaeological record in other places around the world, the role they played on Rapa Nui was not well understood until discoveries in recent years revealed just how extensive the use of both was on the island.

Manavai are relatively small, usually circular, rock-wall enclosures.[10] Some of them stand six or more feet high while some are only one foot or so high, and others are underground. In some cases, the walls of *manavai* are constructed masonry-style, with rocks stacked and fitted atop each other in a single layer. In others, walls are constructed with well-defined parallel rows of boulders placed about three feet apart and with gravel fill set in between the rows. The walls might even be just piles of boulders. *Manavai* may be either singular structures or constructed in a honeycomb fashion. From accounts of their use elsewhere, we know that they facilitated growing crops like bananas, taro, and sugarcane, as well as paper mulberry, used to make bark cloth. Their enclosing walls protect plants from winds, minimizing dehydration, thus helping to optimize the use of available water. They continue to be used by some cultivators today and you can readily see the benefits: the portion of plants that are above the walls of the *manavai* are often brown and torn, while those below are green and healthy.

Manavai also allowed the soil within the walls to be enriched through the addition of household waste and garbage. Our excavations of a few *manavai* scattered along the northwest coast of Rapa Nui have shown soil that is relatively rich in burned material and organics relative to the surrounding earth. Inside the *manavai,* nutrient concentrations, particularly phosphorus and potassium,

Figure 3.2. A walled garden feature, or *manavai*.

are much higher than from soils measured outside the *manavai*. We found this pattern at all of the *manavai* we examined, with the concentrations often two or three times as great. This evidence is consistent with observations in 1786 by La Pérouse, who noted "the natives collect the grass and other vegetables, which they heap together and burn for the sake of ashes, as a manure."[11]

We wanted to determine, at least roughly, the number and distribution of *manavai* across the island, and to do so we were able to harness the power of high-resolution satellite images. While remote-sensing studies cannot replace ground-based investigations, they form an integral part of our research because they allow us to study the entire island through a single image and at relatively low cost. Satellite images are a great first step for documenting prehistoric landscapes, which fieldwork can then study in more detail.

One of our graduate students, Ileana Bradford, took on the project of mapping the location of *manavai* as part of her graduate research at California State University Long Beach. She used three sets of images collected in different years and during different seasons, because under different lighting conditions sometimes what had looked like *manavai* turned out not to be, and at others, some were revealed that weren't visible before. Her careful process provided us with a good overall estimate of their number. A total of 2,553 were identified, with most located within half a mile from the coast, and they are found over large areas of the island.

Of course, the number and distribution of stone enclosures that we observe today is only an estimate of what we might have found at any given point in prehistory. Many *manavai* still used today are likely prehistoric in origin. During our follow-up field surveys, we routinely found thriving banana, taro, and other plants growing in *manavai,* so today's islanders clearly understand how effective they are for cultivation. But we also know that some stone enclosures have been constructed or reconstructed recently, as illustrations of the prehistoric gardening practices for today's tourists.

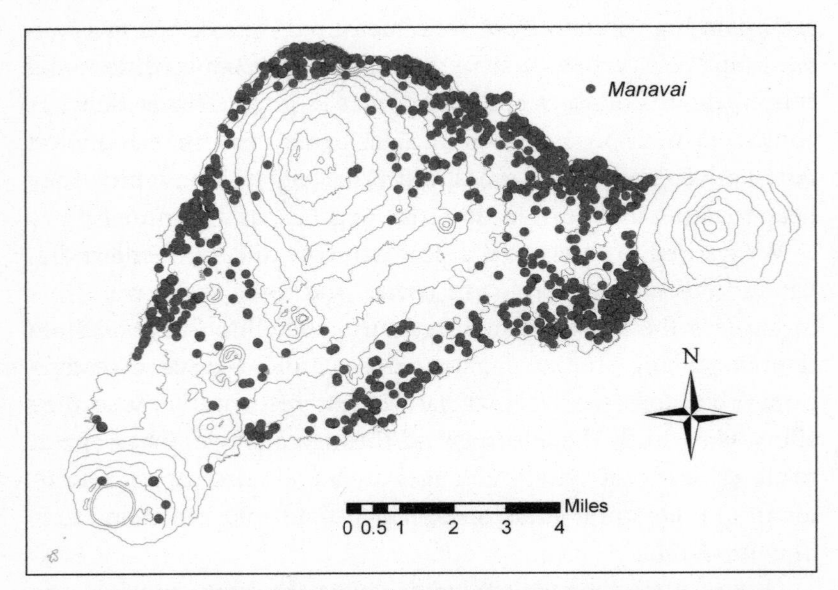

Figure 3.3. Distribution of circular stone enclosures (*manavai*) across Rapa Nui as identified on satellite images.

We would expect that the number would have grown over time, while at the same time, many *manavai* may have been destroyed to make way for other uses or refashioned into other structures. For example, we expect that some *manavai* were dismantled during the nineteenth and early twentieth centuries when so much of the island was converted into a sheep ranch. To control sheep grazing, ranchers constructed extensive stacked-stone structures known as *pirca* walls that stretched across the island, and we know that the *pirca* walls were formed from stones taken from nearby. Some of them undoubtedly came from *manavai,* as well as from other ancient structures. So the number and location of the enclosures, as we have mapped them, likely reflects some large, but not complete, remnant of what existed in prehistory. We expect there are fewer today than in the past because so much terrain is now taken over by the town of Hanga Roa, massive airport runway construction, modern farming, and extensive planting of eucalyptus trees.

Whatever the exact number of them was in earlier times, we could say that taken as a whole, they formed a substantial area for cultivation. The entire area enclosed by *manavai* today is roughly 6.4 square miles, more than 10 percent of the entire island's total surface. This total is even more impressive if one considers that a significant portion of the island consists of the crater lakes, as well as slopes on the shores of those lakes, which are too steep or too rocky for cultivation. With this understanding of the likely extent of *manavai* use, we can certainly conclude that they formed an integral part of ancient farming and that islanders understood very well their critical role in increasing crop yields.

There is an interesting question, though, about why we didn't find signs of *manavai* on various other parts of the island in addition to the shores of the lakes. The most notable area lacking them is the ancient volcano called Poike, at the easternmost part of the island. No stone enclosures are found on the broad slopes of the volcano. This might be explained by geological reasons, as the area lacked the number and size of stones necessary for making *manavai.* Of course, people could have transported rocks to the area, but apparently they didn't. We wondered whether another form of cultivation had been practiced there, but was of limited success.[12]

Early European accounts hint that other types of cultivation were in fact practiced on the island. La Pérouse in 1786, for example, mentions plantations of yams and potatoes as well as banana trees aligned in rows. The captain of the French expedition's sister ship *Astrolabe,* Paul Antoine Fleuriot de Langle, was sent by La Pérouse to explore inland areas of the island. On his trek past the crater at Rano Kau and toward the south coast, Fleuriot de Langle noted that "the cleared grounds have the form of a regular long square, but without any kind of enclosure."[13] Maps published with the written account of La Pérouse show areas on the western coast of the island covered with neatly delineated rectangular fields. A British botanist, Hugh Cuming, described similar cultivation features. During his visit to the island on the schooner *Discoverer* in November 1827, he found that "the Island is . . . extremely well cultivated the ground being laid out in square patches and those close to each other gives it a pretty appearance. Yams, Sweet potatoes, Plantains, Sugar Cane and Coco appear to be principally Cultivated."[14] These observations seem to confirm that the islanders had created cultivated gardens, even though little evidence of field systems has been described in archaeological survey reports.

The first evidence for solving this apparent puzzle came in 1996, when then graduate student in archaeology Joan Wozniak conducted field research on the island's ancient cultivation. Her study area focused on a 0.3 by 0.6 mile portion of the northwestern coast of the island known as Te Niu. Carrying out a systematic survey, she walked across the landscape in a series of transects spaced every fifteen feet, recording all artifacts and architectural remains, as well as conducting small excavations. She found no obvious evidence of cultivation, encountering nothing but rock fields. But curiously, beneath the surface, buried in the soil, she found broken rocks, pits, and artifacts such as obsidian flakes. Thus, while the surface remains appeared to be just a carpet of cobbles and boulders, the subsurface demonstrated plenty of prehistoric activity.

This finding was perplexing at first, but then two different sources of inspiration led her to realize that these scatters of rocks were actually related to cultivation. First, while on Rapa Nui, Wozniak was shown by a local, Niko Haoa, how he protected taro that was growing in his gardens by placing rocks around the plant. Then, in reading the account written by the French explorer La Pérouse, she noted that on the west shore of the island he saw "large stones lying on the surface. These stones, which were found very troublesome in walking, are a real benefit to the soil, because they preserve the coolness and humidity of the earth, and in part supply the salutary shade of the trees, which the inhabitants have had the imprudence to cut down, no doubt at some very distant period."[15] The thought hit her in a flash: Could the rocky landscape of Rapa Nui actually represent a human-engineered landscape constructed for growing plants?

Though initially some of her professors dismissed the idea, she forged ahead in investigating it, and her subsequent excavations documented that the surface rocks and the soil underneath the rocks were both substantially modified by humans. Her analysis shows that prehistoric farmers must have placed the surface rocks there. Wozniak's geomorphology professor conceded that the composition of the soils underneath the surface rocks indicated that they must have been enriched by human intervention. Slowly others began to accept that she was on to something. It seemed that at Te Niu the islanders had practiced a technique known as lithic mulching.

The signs of lithic mulching have been found in excavations of ancient cultures all around the world. One well-known example is that of the ancient Hawaiians on the Kona Coast of the island of Hawaii. Here lithic mulching takes the form of great alignments of volcanic rock and large piles of stones in which a variety of crops were grown. Other locations with remains of lithic mulching include stone mounds in the Negev Desert of Israel, the pebble-mulched fields of Lanzhou, China, the ash fields of the Canary Islands, the rock mounds of prehistoric Hohokam in Arizona, and the pebble-mulched fields of prehistoric Anasazi in

New Mexico. Farmers in New York and New Jersey have also used lithic mulching as recently as the 1930s and 1940s. Farmers in northern Ohio as well as gardeners in New York City also practiced lithic mulching in the 1960s and 1970s.

Lithic mulching increases agricultural productivity in several ways. First, the surface rocks protect plants by generating more turbulent airflow over the garden surface. This results in a reduction of the highest daytime temperatures and an increase in the lowest nighttime temperatures, which produces a healthier growing environment for plants. In addition, the disrupted airflow limits the amount of wind that batters the foliage, similar to the protection offered by the walls of *manavai*. The placement of rocks, particularly broken, smaller ones, serves another essential function: it increases the productivity of the soil by exposing fresh, unweathered surfaces, thus releasing mineral nutrients held within the rock. By breaking down large rocks into small pieces, one can maximize the exposed surface area available for mineral leaching. Relative to a single large boulder, many fist-sized rocks of the same total volume have many times the amount of surface area. Often the rocks are placed not only on the surface, but also buried to directly introduce new sources of minerals into the soil.

Despite growing acceptance of the notion that the Rapanui had made use of the practice, there was continued skepticism because it suggested such a different understanding of the culture and history of the island. While Wozniak's research itself was widely accepted, many resisted the greater implications, since they directly upend the long-standing and oft-repeated belief that the rock-strewn landscape is unproductive, and was degraded as the result of the "imprudence of the ancestors," to quote La Pérouse. But over time, the idea began to take firm hold. The late Roger Green, a well-known Pacific archaeologist, began to wonder whether he had seen the same kinds of lithic mulch in areas where he had worked, such as in Hawaii. Archaeologists Chris Stevenson and Sonia Haoa found the same kind of patterns of surface rock and modified soils in excavations they conducted along the northeast coast of Rapa Nui.[16] The evidence began to grow.

We also found that the implications of her work took time to sink in. During the first several years of our fieldwork, we commonly commented to each other about the incredibly rocky terrain that makes up most of the island. Faced with guiding field school students on foot surveys across the landscape, we were always concerned with twisted ankles—injuries that are relatively minor in most places but worrisome on this remote island. The undeveloped land of Rapa Nui is literally a minefield of ankle-twisting rocks. Not only does the high density of rocks make the surface a hazard, but also the sizes of them seem almost designed to cause a tumble.

Throughout our surveys, we often cursed and pondered these swaths of billiard balls. Over time, however, we came to recognize that these dense patches of stone were located over a remarkable quantity of the island. We found them in flat areas, at the bottom of the slopes, on hill slopes, and in swales. We discussed this endlessly as we walked our survey transects and wondered what kind of geological or erosional process would result in this pattern. Are

Figure 3.4. Lithic mulch garden near Ahu Akahanga on the south coast of Rapa Nui.

Figure 3.5. Taro growing in lithic mulch, Te Niu area, on the northwest coast of Rapa Nui.

the rocks rolling down hills and accumulating after being exposed through erosion? No—there are no such rock exposures above the rock fields. Are the rock scatters caused by sheep that once covered the island by the tens of thousands? The explanation continued to elude us.

We had read Wozniak's work and we understood that she had identified features on the northwest coast that she called "rock gardens," but we had the impression that these must be fairly limited in size, gardens such as we think of them today, not vast expanses of stone. The mental leap we had to make was to see an entire landscape engineered, in a sense, as a garden.

A couple of bits of critical information came together to fully open our eyes to what we were seeing. First, other researchers began to find indications of these rock gardens in more and more locations across the island.[17] Eventually, German researchers Hans-Rudolf Bork, Andreas Mieth, and Bernd Tschochner calculated that stone gardening activity could be found across an area that spans almost one-half of the island's surface. Then, perhaps even more important, archaeologist Thegn Ladefoged with colleagues Chris Stevenson, Peter Vitousek, and Oliver Chadwick published a paper that showed that the island's soils had remarkably poor mineral content.[18] Based on chemical studies of sediment derived from the slopes of two of the island's volcanoes, Terevaka and Rano Aroi, where the most mineral-rich soil would be expected, they learned that phosphorus, which is important for plant growth, is uniformly low. Their work showed that unlike other volcanic islands in the Pacific, Rapa Nui has remarkably unproductive soils and, critically, that they have always been unproductive.

This point is essential to understanding the prehistory of Rapa Nui. While countless scholars have commented on the current poor condition of the environment, we now know that this situation was in existence long before the arrival of humans. Even when the island featured a palm forest, the soils were not particularly fertile.[19] While historic erosion from sheep ranching in the last century may have left the island with even less fertile soil, the contemporary environment is not much different from what prehistoric occupants faced in their struggle to grow crops.

The picture that had emerged provided an entirely new understanding of the prehistoric record. We had long assumed, as had many others, that Rapa Nui's volcanic origin bestowed it with reasonably productive soils. The largest constraint on agriculture on the island, we had assumed, was lack of reliable rainfall and flowing streams that could have been diverted for irrigation. But now it was becoming clear that Rapa Nui's soils had been fundamentally unproductive, and for a very long time. According to recent models[20] of the volcanic origins of the island, the bulk of Rano Kau on the southwest corner of the island was formed by eruptions that occurred between 450,000 and 940,000 years ago. Terevaka, to the north, and the source of much of the island's overall landmass, was volcanically active between 460,000 and 780,000 years ago. Overall, these volcanoes are old enough to have lost primarily mineral nutrients.[21]

Initially, the soil that is formed from freshly erupted volcanic ash and rock contains abundant minerals. Phosphorus and nitrogen, vital parts of photosynthesis—the conversion of solar energy to chemical energy—are abundant enough. Over time, however, the quotient of these nutrients declines with leaching from rainwater and use by plants. Consequently, while young volcanic islands are some of the most biologically productive places on earth, those with older volcanoes can be impoverished, even with adequate rainfall. In fact, abundant rainfall exacerbates the situation as mineral nutrients are flushed from the soil.

In the case of Rapa Nui, the volcanic soils are hundreds of thousands of years old and greatly depleted of their nutrients. The island, therefore, has been a poor place in which to make a living by farming since long before people arrived in AD 1200. Indeed, studies conducted by soil scientists Geertrui Louwagie and Roger Langohr confirm that water was not the main problem for the prehistoric farmers on the island, but rather, limited soil nutrients were. Using data for crop growth coupled with experiments of cultivation in four areas on Rapa Nui, Louwagie and Langohr demonstrated that only the addition of lithic mulching made soils rich enough to support even marginal conditions for plant growth.

This understanding was a revelation to us. The truth of cultivation on the island was that only the ingenuity of the islanders made it possible to produce a reliable food crop. One immediately obvious implication of lithic mulching as a central part of subsistence on Rapa Nui is that there must have been a staggering amount of labor invested in moving rocks. With an estimate of thirty square miles, the number of rocks that prehistoric islanders moved, broke, buried, and scattered on the surface is astronomical. Indeed, based on a study of lithic mulching at more than five hundred sites across Rapa Nui, Bork and his colleagues estimate that the total amount of stones weighed in excess of two million tons and individually numbered well over a billion. Given that many of the rock sources are nearby, but still a short distance away from the cultivation areas (about 150–200 feet), they estimate that the islanders traveled an aggregate of eight million miles over the duration of five hundred years of prehistoric cultivation.[22]

With this new understanding of the intensive work the islanders put into making cultivation possible on the island, we began to revisit the issue of the islanders' role in the fate of the *Jubaea* palm forest. The evidence was strong that it was a boom in the population of rats that had contributed significantly to the loss of forest. But we might ask, why didn't the islanders work to replenish the forest? If they were such dedicated stewards of their environment, this might have been expected. Indeed, while people probably did not engage in wholesale destruction of the forest, they did clear some forest.

Part of the answer comes from the fact that the *Jubaea* palms are slow growing. It takes several years for the tree to form a trunk and sixty years or more to produce seeds.[23] With rats consuming so many of the palm nuts, as the record suggests, few trees would regrow naturally. Even if the islanders had vigorously planted nuts in an attempt to replenish the forest, the palms would have taken multiple human generations to produce food. Meanwhile, rats would happily eat tender young palm seedlings as well as the nuts. As long as there were palms, there would be nuts, and rats to feed on them. Cultivating land on which the forest once stood meant a higher yield of food in Rapa Nui subsistence. As palms declined, more area became available for planting, and burned organic material from palms would have provided an important, albeit temporary, source of nutrients for crops. In this way, the decimation of the forest was by no means an ecological disaster, as least not as far as the human population was concerned.

This is the form of cultivation popularly known as slash-and-burn. Forms of it have been used in nearly every forested environment, including other Pacific islands, northern Europe, the Amazonian rain forest, Southeast Asia, and prehistoric North America. Also aptly called "shifting cultivation," the strategy typically requires populations to move from place to place, as the gain in soil productivity is temporary. When populations are able to move garden plots, slash-and-burn strategies can be sustained for long periods of time. Often groups follow a long-term rotation of land use, returning to areas only after trees and soil nutrients have regenerated.

On Rapa Nui, of course, neither expansion to new forested areas nor long-term rounds were possible within a short time, so shifting cultivation was only a temporary solution. Consequently, farmers soon switched to intensive use of *manavai* and lithic mulch.

Now let's revisit the question of why we find no *manavai* or rock mulch gardens on the slopes of Poike, the easternmost volcano on the island. Poike is distinct because unlike the slopes of the other two volcanoes, Rano Kau and Terevaka, its slopes contain almost no rock. Instead the slopes of Poike are covered with fine-grained soil from volcanic ash.

The lack of surface rock played a central role in traditional accounts of the conflict between two groups of islanders known as the Long Ears and Short Ears, said to have escalated out of control. Thor Heyerdahl, for example, writes:

> The long ears' last idea was to rid the whole of Easter Island of superfluous stone, so that all the earth could be cultivated. This work was begun on the Poike plateau, the easternmost part of the island, and the short ears had to carry every single loose stone to the edge of the cliff and fling it into the sea. This is why there is not a single loose stone on the grassy peninsula of Poike today, while the rest of the island is thickly covered with black and red scree and lava blocks.
>
> Now things were going too far for the short ears. They were tired of carrying stones for the long ears. They decided on war. The long ears fled from every other part of the island and established themselves at the easternmost end, on the cleared Poike peninsula.[24]

We would assert that this is a rather far-fetched tale. From the perspective of European farming, Poike would seem to be a superior location for growing crops, and according to that view of farming, rocks must be removed to make cultivation possible, especially with plowing. However, the volcano is ancient, even more so than

Terevaka, with the ash deposits that form its slopes produced by eruptions some 400,000 to almost 900,000 years ago. Perhaps even more than the rest of the island, the volcanic ash soil of Poike is heavily weathered and, as a result, is poor in mineral nutrients. These poor soils would pose a great challenge in producing any appreciable crop yield. Archaeological evidence shows that early colonists did cultivate the soils of Poike. But the evidence also indicates that these efforts at cultivation were soon abandoned.[25]

Using satellite images for our research, we were intrigued to find large swaths of parallel lines that showed up in our photos on the slopes of Poike. These lines are clearly visible in the images available on Google Earth. These parallel lines bear the unmistakable characteristics of crop furrows, marks made by the farmers to prepare soil for crops. We showed these marks to Sergio Rapu, a local archaeologist who earned a master's degree in anthropology at the University of Hawaii and served as the first native Rapanui governor of the island, and wondered whether this was previously unknown evidence of ancient farming. Sergio chuckled and immediately recognized the marks as an aborted attempt to grow corn on Poike just a decade or two ago. Despite the availability of industrial tools, the poor quality of the soil does not support crop growth.

So if the islanders wanted to cultivate the volcano's slopes, they would have needed to enrich the soil. Lithic mulching might have been a way to do so, which would, of course, have meant they would have brought rocks in to cover more of the surface, not have taken them away. But as we've seen, lithic mulching only marginally increases the quality of soil, and so it makes sense that the islanders would not have engaged in the practice at Poike.

We do see evidence of soil erosion on the volcano, and this has been cited as evidence for prehistoric environmental degradation. Indeed, it is clear that a palm forest once existed on Poike, and it is likely that it suffered the same fate as the forest elsewhere, being depleted both by rats and by the clearing of the land in initial attempts at cultivation. At first, burning the palms and other vegetation would have enriched the soil enough for crop growth, but soon after, cultivation was no longer feasible. No dramatic story of overexploitation or ecological collapse is needed to explain this outcome.

Our new understanding of the strategies used by the islanders to produce food in a sustainable fashion highlights the relationship of the human population to the environment of Rapa Nui. We now know that the island was never particularly productive, given the limited marine resources, the small number of introduced animals and cultigens, and the nutrient-poor volcanic soils.

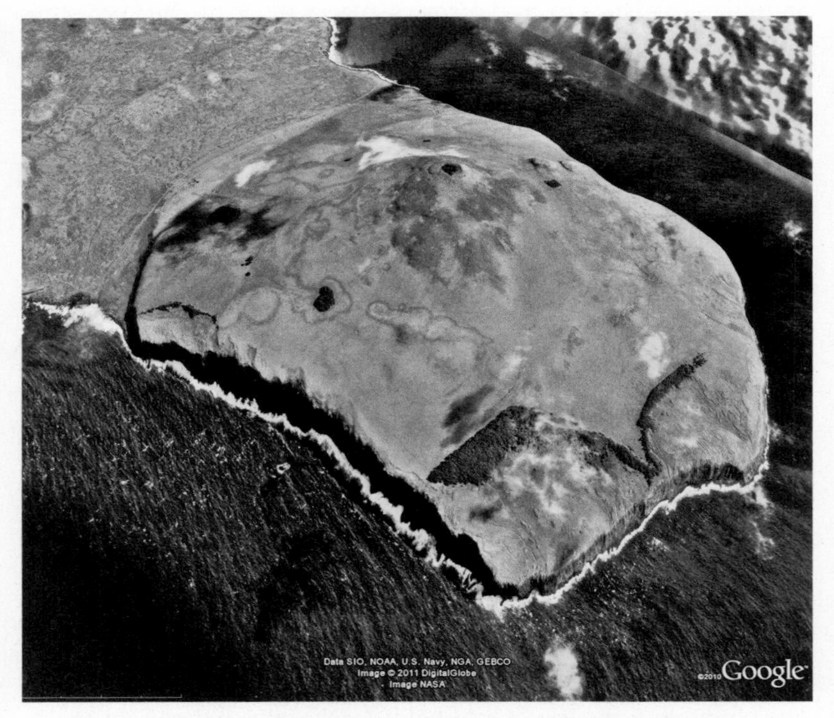

Figure 3.6. Google Earth perspective of the Poike Peninsula. The large scarred areas on the south and southeast margins of Poike are the result of historic erosion from sheep ranching. The lines along the eroded areas are eucalyptus trees planted in attempts to stem further erosion. These conservation efforts have not been particularly successful.

Over the long run, the islanders invested their energy in effective efforts to produce food in good times and bad given the resources available to them.

In light of this knowledge, we can readily see the unwarranted nature of claims for a prehistoric environmental catastrophe that turned a once-productive island into a barren landscape. If anything, the islanders contributed to an increase in the human carrying capacity of the island over time. We can also readily see that there is no reason to suspect that population sizes for the island ever greatly exceeded the numbers witnessed at the time of European contact in AD 1722. The first Europeans encountered a functional economic and social system shaped by five hundred years of experience of making a living on this modest island. The population of around three thousand recorded in 1722 reflects a sustainable size for the island, not one dramatically reduced through conflict and starvation.[26]

All of these findings suggest that rather than a case of abject failure, Rapa Nui is an unlikely story of success. Using the skills, knowledge, and materials available, and adapting them to meet the specific conditions, the islanders transformed Rapa Nui from an island covered in palm forest, with few resources for humans, into an island that could reliably, though marginally, sustain them over the long run.

Initially, the islanders practiced slash-and-burn cultivation, and as the forest declined, they created a series of *manavai* gardens while also laboriously turning the landscape into an engineered series of massive fields fertilized by broken volcanic rocks placed on the surface and in the ground. Little by little the island was transformed into an endless series of gardens. The story of Rapa Nui is one not of ecological suicide but of persistence and resilience in which the islanders employed innovative approaches and a willingness to invest massive amounts of labor.

Our understanding of the history of ecological management of Rapa Nui also contradicts, of course, the notion that the forest was depleted in large part for the purpose of building contraptions for transporting the massive statues. But if this wasn't the case, then the question of how the islanders did manage to move their statues becomes all the more puzzling. And if the islanders weren't the rapacious destroyers of their environment they have been depicted as, then we must revisit the notion that making and moving statues became a great burden to the island and its culture, contributing to its collapse. So let us now turn to this part of the Easter Island mystery.

CHAPTER 4

The Ancient Paths of Stone Giants

> In Easter Island the past is the present, it is impossible to escape from it. . . . The shadows of the departed builders still possess the land. Voluntarily or involuntarily the sojourner must hold commune with those old workers; the whole air vibrates with a vast purpose and energy which has been and is no more.
>
> —Katherine Routledge,
> *The Mystery of Easter Island,* 1919

Perhaps the biggest mystery surrounding Easter Island is the question of how these enormous statues were moved. Because we find a host of statues in the quarry at Rano Raraku, some abandoned in the process of being carved, we know that the ancient islanders transported the multi-ton carved statues from the quarry to their current locations across miles of the island's rugged terrain. Early visitors were perplexed by such an astonishing feat, given the lack of trees on the island for providing the timber and rope that would presumably have been needed to make devices for moving them. What's more, the island lacked draft animals, and even the technology of the wheel, which would seem vital for whatever type of transport vehicle they made. The islanders had moved more megaliths over greater distances with fewer people and resources than any ancient society known. No visitors ever witnessed their transport, nor do any records document it. But everyone has a theory.

The natives' own folklore, handed down from generation to generation, tells of chiefs and priests imbued with mana (the concept of supernatural power found throughout Polynesia) who simply ordered the statues to walk. The statues walked for a distance each day, then stopped. Day after day the priests would invoke the same rituals until the colossal *moai* had reached their destinations.

In the search for scientific answers, one assertion that had been made through the years was that the islanders had built roads from the quarry for transporting the statues, but this was not well established. When we began our work on the island, we did not expect to be drawn into this mystery of the statue transport, but a stop we made one day in 2003 as we introduced a new set of students to the island piqued our curiosity.

As part of our archaeological field schools, we have a tradition of introducing students to the island through field trips the day after they arrive. The tour consists of a "best of" list of sights and generally follows a route taken by many tour groups. There is a particularly dramatic spot at which we pause when the road crests a broad ridge, providing a view of rock- and grass-covered landscape that stretches from Poike on the eastern tip to the Rano Kau volcano to the west. As we approach the coast, we begin to see heaps of rock at regular intervals, which on closer inspection are seen to be the remains of the large stone platforms on which the islanders mounted their statues.

As students look closer they suddenly spot the *moai* that lay toppled in front of the *ahu*. On the road we pass one *ahu,* then another, and another, and another. Most of these large platforms have a name passed down by tradition—Ahu Hanga Te'e o Vaihu, Ahu Akahanga, Ahu Ura Uranga Te Mahina, Ahu Oroi, Ahu Tongariki, and many more. As we drive down the rough two-lane road along the rocky shore, each bend in the road reveals more. The abundance and scale of the archaeology one sees along the south coast of Rapa Nui is overwhelming. While this might sound like hyperbole, it is hard to grasp the idea that an island this remote and tiny would have so many prehistoric monuments that are so massive. Almost without exception the otherwise boisterous students become quiet and contemplative as they process what they are seeing. While many of them have previously seen and even worked at impressive archaeological sites, the views along the coast are jaw-dropping. What they are viewing is not just a single location but a landscape of prehistoric monuments and archaeological remains: a landscape of massive statues and platforms stretching as far as one can see down the coastline.

For years, on our field trips with students we had passed a decrepit, half-standing wooden sign with a bilingual message carved into it and still barely readable: "El Camino de los Moai" and in the native Polynesian language Rapanui, "Te Ara o te Moai." The sign, which is no longer present, seemed to point to an eroded rut in the barren rocky and red dirt landscape. Was this sign really pointing to a road for transporting the ancient statues? We were skeptical. We had many reasons not to take the sign seriously. Speculation about how the statues were moved has gone on for decades, and no one has made a big deal about there being roads on which they were transported. If there has been evidence for such roads, why have people gone so far as to suggest such wacky ideas as that the statues were shot out of volcanoes and landed where they are now, or that alien technology was used to transport them through the sky, or that they were floated on rafts of palm logs? Surely, we thought, the sign was made for the amusement of tourists. There are certainly a lot of little signs and plaques like this on the island, indicating spots for tour buses to stop and let their passengers snap photos.

Yet after stopping our caravan of jeeps to get out and let the students stretch their legs that day, we decided to follow this unremarkable linear depression a few hundred feet up into the grass. And in the middle of the track we encountered a multi-ton statue resting on its belly, with its head pointed downward and toward the ocean. We were still new to the island, taking in as much as we could to understand the archaeological record, and this perplexed us. That such an ancient earthen road would still be visible today seemed hard to believe. How could it have survived all these years? And if it really was such a road, why weren't these roads more famous?

Part of our disbelief about the roads was due to the fact that we had consciously decided not to engage in the controversy over the moving of the *moai*. In our initial field seasons we had even avoided close examination of the statues themselves. Despite their fame, the tremendous amount of ink dedicated to the *moai* and theories about the technology used to make and move them led us to focus on other issues. The topic seemed too fraught with plausible alternatives, each without a clear means of falsification. There were debates that would never be resolved, we reckoned. We also assumed that many of the issues related to *moai* manufacture were well established. In fact, as we said before, we began our work with the assumption that the general outline of prehistory for the island was pretty well-known. So, for some time, we did nothing to follow up about the issue of the roads. But as we continued with our field research, surveying on the south coast of the island in an area called Akahanga, we began to find more evidence that there had in fact been *moai* roads.

Surveying inland from the coast, we began to see intermittent stretches of rock in parallel alignments about 10 or 12 feet apart extending over the rolling landscape and often enclosing eroded earth forming a rut. Sometimes parallel lines of stone enclosed flat ground, often remarkably free of stones. At first we thought these features must be related to the extensive sheep ranching from the late nineteenth and early twentieth centuries. But then we realized that they tended to disappear beneath the historic walls used in sheep ranching, reappearing on the other side, so they must have predated the ranch walls. We finally came to the conclusion that these were, in fact, fragments of El Camino de los Moai.

We started to wonder whether there were artifacts of such roads elsewhere on the island. Where did they go and how extensive a network had they constituted? Would there be a way to find direct evidence that they were used to transport the statues? And if so, might they also offer clues about the manner in which the giant monuments were moved?

As we delved into the available literature about the roads, we found that there was, in fact, a long history of observation of them. The first mention of the roads in the records about the island was made in the account written by the naturalist who accompanied Captain Cook on his visit to the island, Georg Forster. One group of the English visitors traveled eastward across the island. As described by Forster, the party followed a "path" that led them through a landscape covered in stones and with gardens of sweet potatoes. Along this path they encountered *moai,* some of which stood on stone platforms. In addition, Forster described fallen statues, as well as large standing statues situated along the path. Forster states that along the south coast "they observed that this side of the island was full of those gigantic statues so often mentioned: some placed in groups on platforms of masonry; others single, fixed only in the earth, and not that deep; and these latter are, in general, much larger than the others."[1] Given that they were not standing on platforms, these *moai* are likely ones that were in the process of transport.

The next written account that appears to mention *moai* roads is that of Lieutenant Captain Wilhelm Geiseler of the German Imperial Navy, who visited the island in 1882.[2] Geiseler commanded the German battleship *Hyäne* and was ordered by the Berlin Imperial Museum to conduct an ethnographic study of Rapa Nui. During his four-day stay, his group traveled along a *moai* road on the south coast "a few hundred meters back from the beach," which he wrote stretched from "Waihu to Rana Roraka," presumably referring to Vaihu and Rano Raraku. Along this trail, Geiseler describes about twenty fallen *moai.*

But by far the most extensive record of observation of *moai* roads comes from Katherine Routledge's observations in 1914–15. Routledge had set up camp near the local sheep ranch manager's office and began to explore the island. Living in tents, she wrote, was not "beer and skittles," given the nearly constant wind.

Routledge described everything she could about the island:

the archaeological record, traditional stories, language, and the customs of the people. Using a local native interpreter, she recorded oral traditions and ethnographic information, including as many of the stories about the statues as she could. Much of the value of her work is in these accounts of the people of Rapa Nui, who, at the time of her visit, numbered only about 250 individuals in the wake of disease epidemics and slave raids. Soon after Routledge's arrival on the island in 1914, tensions between the native Rapanui and foreign managers of the ranch escalated. The Rapanui stole cattle and feasted on it in an act of rebellion, at the same time declaring war on the ranch. Much unrest surrounded the harsh conditions the Rapanui confronted. As Routledge describes the situation, "the statues remained quiescent, the natives did not."[3]

Routledge compiled a systematic inventory of *moai* and *ahu* of the island as well as undertaking excavations, and in her 1919 book, *The Mystery of Easter Island,* she recounts an event that led her to notice that the *moai* were aligned along what looked like paths. While sitting on the rocky slope of the statue quarry at Rano Raraku, she looked west along the south coast of the island. In the late afternoon light, she noted that "the level rays of the sinking sun showed up inequalities of the ground, and, looking towards the sea, along the level plain of the south coast, the old track was clearly seen; it was slightly raised over lower ground and depressed somewhat through higher, and along it every few hundred yards lay a statue."[4] One can make the same observation today visiting the quarry. Perched on the edge of the crater of Rano Raraku, looking across the rocky south coast of the island, one sees an unmistakable line that angles away from the quarry and along which are dozens of fallen statues.

Beginning with this observation, Routledge discovered an arrangement of such roads over the island and sketched a map showing segments of them (Figures 4.2 and 4.3, pp. 62 and 63). Her work resulted in the identification of at least seven road fragments or segments.[5] Routledge's map has remained the primary

Figure 4.1. A view of one of the *moai* roads leading from the quarry at Rano Raraku to *ahu* located along the south coast of the island. This position is probably close to where Katherine Routledge made her initial observations of the roads in 1914. The dotted line traces the path of an ancient road and the arrows point to locations where there are fallen *moai*.

documentation of these roads, and surprisingly little additional research had been done on them until recently.

Following Routledge, Thor Heyerdahl did the next important work on the roads. In his fieldwork, carried out in seasons over 1955 to 1965, and then again in 1986, Heyerdahl was particularly interested in explaining how the Rapanui had managed to move their massive statues. To investigate this mystery, his team excavated the area around a fallen *moai* located on a prehistoric roadway. They uncovered "a densely packed surface of broken and crushed lava" just below the surface,[6] and Heyerdahl interpreted this surface as evidence for hard-packed and unpaved roads across which *moai* were transported. This roadway was "slightly wider

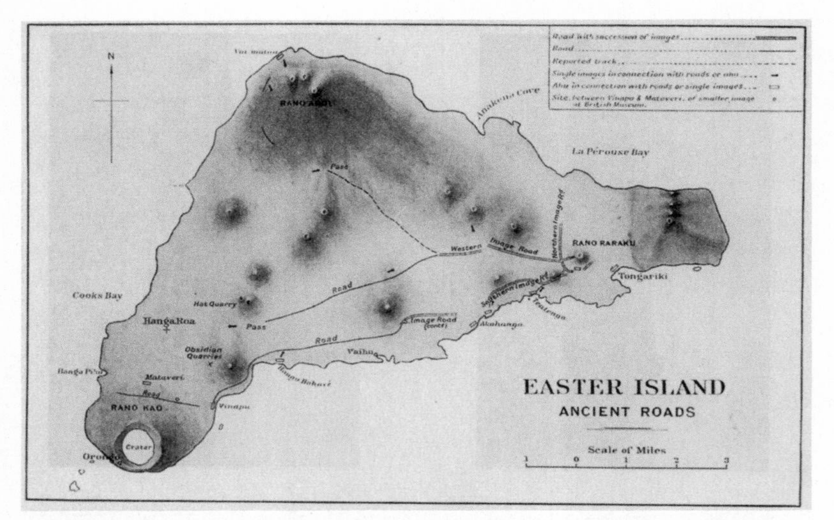

Figure 4.2. Map of ancient roads documented by Katherine Routledge during her fieldwork on the island in 1914–15.

than the [*moai*] base [and] was as firm and even as if a steamroller had prepared the ground."[7] The hard-packed surface was, according to Heyerdahl, the result of the movement of massive statues across the ground, a record of their tracks. The team also uncovered a large number of stone hand axes (*toki*) arranged in a circle, where an adjacent statue had once stood. Heyerdahl conjectured that these hand axes had been wedged under the edges of the statue when it was standing, but he made no further speculation about how they might have been involved in either the making or the transporting of the statues.

Charlie Love, a geologist cum archaeologist, has undertaken the most intensive research on these ancient roadways and their features.[8] Love is a regular fixture on Rapa Nui and brings years of firsthand experience exploring and working the island's archaeology. After decades of work, he is probably the archaeologist who knows the island best, except, perhaps, Sergio Rapu. Though much of his work remains unpublished, Love is eager to share his recollections and his impressive control of the details of the island's archaeological record. Love teaches in a community college in Wyoming, where he makes up the entire geology department. He is a classic western guy, and the kind of "dirt"-based archaeologist who values the things he has held with his own hands or seen with his own eyes. And he has seen a lot of the island. The papers he delivers at the annual conferences on Rapa Nui archaeology are characteristically full of surprising photos showing corners of the island about which few are aware, historic photos unearthed from his endless scholarship, and detailed comments about the record gained only from decades of patient observation. He is also well-known as the kind of archaeologist whose encounter nearly inevitably results in late-night tales and debates over drinks. In short, he is a classic field archaeologist.

In 2000, Love mapped about five miles of the roadway used to deliver Rano Raraku quarry *moai* to *ahu* along the south coast. Along this stretch, Love noticed that the roadway width is quite consistent, measuring about eighteen feet across, traverses changes in topography, and was constructed to clear the path of

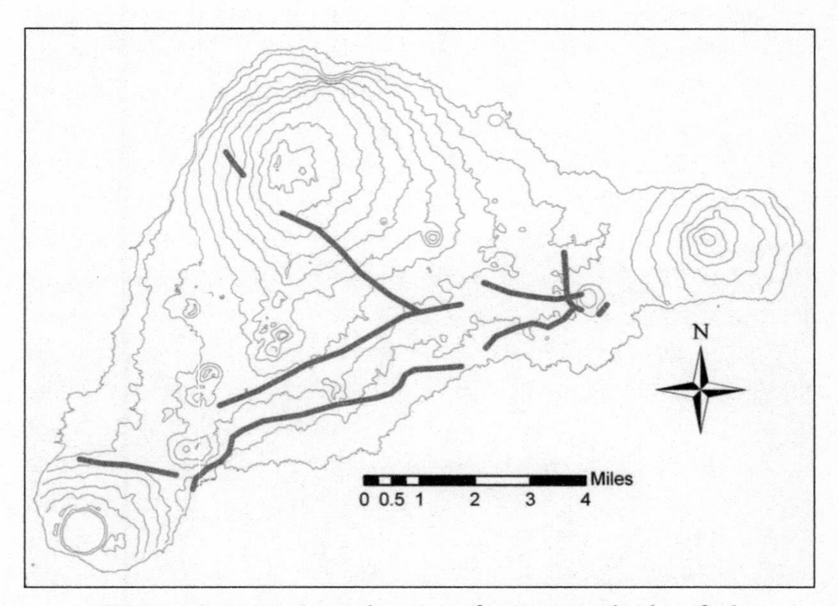

Figure 4.3. Approximate location of ancient roads identified by Katherine Routledge in 1914 shown on a modern map.

obstacles. Love also documented features of the roads that affect the effort expended in moving statues. For example, when these one-way roadways leading out of the quarry go up and over a rise in the landscape, its incline never exceeds three degrees. The minimal amount of slope must reflect the paths chosen to meet the exigencies of transport. Heading downhill, however, roads decline as much as six degrees. These observations suggest that there are different constraints involved in moving a statue uphill versus downhill, and as we shall see later, this issue is helpful in gaining an understanding of just how the statues were moved.

Love's surveys reveal that a good deal of work was put into making the roads in order to create flat paths over the rolling and rocky terrain of the island. When roads traverse up a hill, the roadbed was cut into the land; and often on the downslope, especially in low valleys, the roadbed was raised substantially in order to decrease the steepness of the slope. The material added in these cases often raises the roadbed as much as three feet.

Figure 4.4. *Moai* road and fallen *moai* along the side of the roadbed, south coast of Rapa Nui.

Love also observed that roadways have "curbs" in some places, consisting of aligned sets of boulders. It is possible that these curbs were placed to demarcate the edges of the road, and they might have prevented soil from eroding into the paths. These curbstones are not found consistently along the length of the road, and there is no obvious reason why we find them over some stretches and not others. Love speculates that the curbstones may have been fulcrums for levers that were used to move the statue forward. Love noted as well small piles of large angular boulders near the roadways at intervals. These boulder piles may also have been associated with *moai* transport, but their function remains unknown.

Love also documented water-worn beach boulders, called *poro,* located along the roadways about every one hundred meters. These *poro* stones are distinctive oval-shaped boulders (around 15 to 25 inches in diameter) that are composed of dense grayish basalt with a smooth surface. *Poro* stones were collected by prehistoric Rapa Nui in the tens of thousands and brought inland for use in many types of architecture, such as pavements for *ahu* and the elaborate boat-shaped houses called *hare paenga* or *hare vaka.* What makes the *poro* stones he found along the *moai* roads particularly curious to Love is the fact that they are routinely broken in half or have large chipped sections knocked off them. Oddly, Love found only one half of the broken stones; the other halves of the boulder and any flakes that were driven off in the fracturing process were nowhere to be seen. Given their density and rounded shape, an enormous amount of force must have been involved in their fracture. Their regular placement along the roadways suggests that they were used as part of statue transport, and Love conjectures that the *poro* stones were occasionally crushed in some way by the weight of the statue being moved. It is possible the stones were used as wedges underneath the statue, though no one knows the exact role they played.

In addition to his mapping of these five miles of the south coast roadway, Love has conducted excavations in five other locations along the road. These excavations have exposed more than six

Figure 4.5. *Moai* roadbed along the south coast of Rapa Nui.
On the right-hand side of the road are embedded "curbstones."

hundred feet of road and have illuminated the road composition. Under the surface, Love found layers of densely compacted earth, just as Heyerdahl had, and they appear to represent the original roadway surface. Some of the roads that Love excavated had multiple older surfaces, suggesting that they had been filled in and then resurfaced over time.

Love found from excavating cross sections of road that the surfaces varied from flat to slightly U or V shaped. In other words, some of them were more like grooves than flat roads. Love suggests these roadbed shapes have to do with the shape of the devices that were used in *moai* transportation, but as yet no evidence of any such devices has been found. It might also be the case that the grooved shape is the result of erosion that might have occurred after the road was no longer being used.

A further clue as to how the statues were moved along the roads may come from Love's discovery of a series of relatively small shallow pits or "post holes" located just beyond the edges of the roadway surfaces. These pits vary in size from 6 to 12 inches and from 12 to 24 inches deep. Love suggests that these pits might have been related to statue transport, perhaps serving as the foundation for posts that were used in some way. But again, we don't have any firm evidence that this was the case, and there might be other explanations. For example, the holes might be pits dug for gardening in the back-dirt accumulated from the road clearance. We simply don't know.

Clearly many mysteries remain about the roads, and we grew more and more intrigued by them. We wondered whether the full extent of the roads could perhaps be mapped if we were to once again make use of satellite images. Archaeologists working in other parts of the world had used satellite images to trace ancient roads, even footpaths.[9] The success of these studies made us optimistic. We turned to high-resolution images that we obtained from a company called DigitalGlobe, and sure enough, the minimal vegetation and the nature of the archaeological record on the island made all sorts of features easily visible in the images. We easily identified the linear marks of roads, and we were able to locate many known roads, such as those already identified by Routledge and Love, as well as many more potential roads that had yet to be identified and mapped.

We had to be careful in our designation of any given road as a *moai* road, though. Extensive sheep ranching in the late nineteenth and early twentieth centuries as well as recent farming activities have resulted in a wide array of roads, stone walls, and fences that could be mistaken as parts of the ancient roads. So after initially plotting as many linear features as possible, in order to distinguish the ancient roads we paid particular attention to the lines that were crossed over by other features such as ranch roads and walls. This allowed us to establish that they had come earlier.

Over the course of several field seasons on the island, we traveled with our student survey teams to every location with a potential road. The presence of statues along roadbeds as well as curbstones and the U-shaped depressions served as our primary means to confirm that these were also *moai* roads.

Some of the alignments we traced on the satellite images no

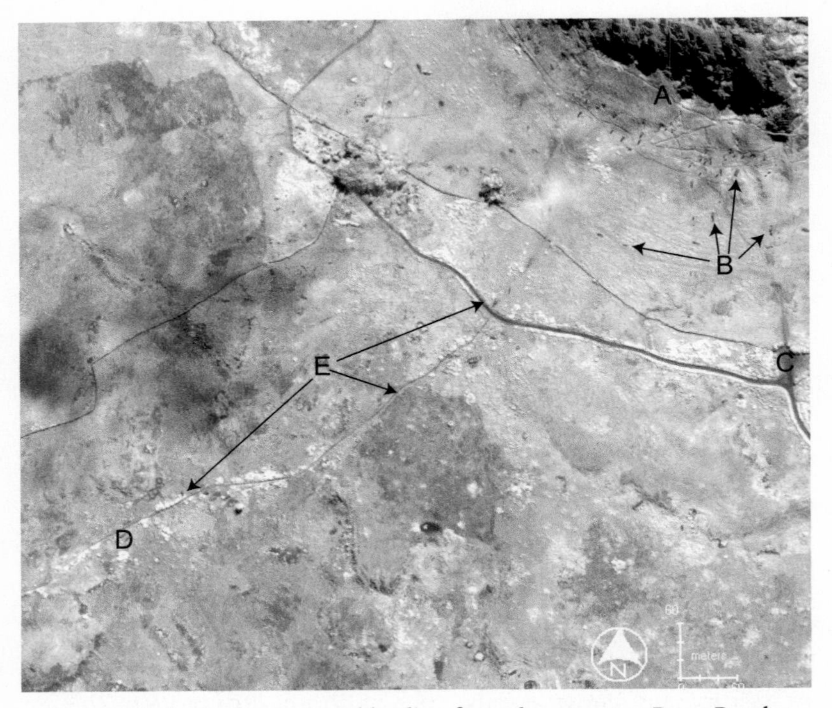

Figure 4.6. Path of a *moai* road leading from the quarry at Rano Raraku in the upper-right-hand corner to the lower left corner of the image. This portion of the road was visible to Katherine Routledge as she observed the area in the afternoon light from the slopes of Rano Raraku. In this satellite image, a number of key features can be identified. (A) Statues that surrount the quarry are easily visible in this image (B), as is the modern parking lot (C). The ancient road (D) is visible primarily as a horse trail and as a line of vegetation that runs from the northeast to the southwest corner of the image. This feature likely reflects sediment compaction with greater water retention and subsequent vegetation growth. Multiple large statues (*moai*) line this road near the quarry (E). The satellite image was provided by RADARSAT, Inc., and DigitalGlobe, Inc.

doubt actually belonged to a later period, while others might include more recent uses of prehistoric pathways. We know this to be true, for example, of sections of the path along the south coast, which are currently covered by a paved road. In another area, the old *moai* road is a popular horse trail. By examining our set of identified *moai* roads closely, we were able to identify some

Figure 4.7. *Moai* roads mapped through satellite imagery and ground survey. The paths of the roads are shown on Google Earth imagery with slight exaggeration of relative elevations to show topography. In the distance to the left is Rano Kau, to the right is Terevaka, the island's highest point, and the crater in the bottom center is Rano Raraku, the statue quarry.

key features that tell us about the history of their construction. Some of the roads were clearly rebuilt several times, as there are slight variations in their paths over time. Significantly, Love noted a similar pattern in his excavations of 2000, when he found evidence that in some places there may have been three separate roadbeds, each built on top of the other.

As we systematically walked these tracks, guided by the satellite images, we discovered statues, one after another, many of which were not yet officially known. Of course the islanders know about these remote *moai,* but they had escaped documentation by archaeologists. In our road surveys, we added about 40 "new" statues to

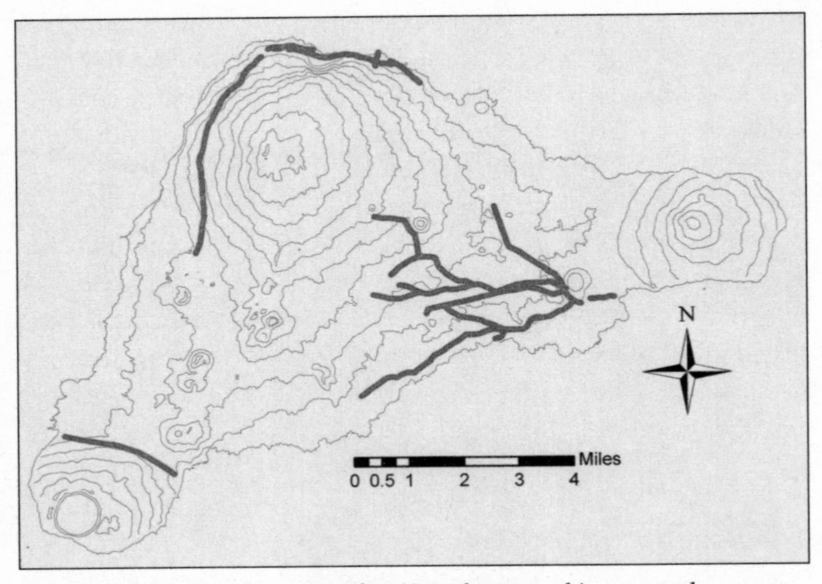

Figure 4.8. Location of *moai* roads mapped in our study of satellite images and ground survey.

our own growing database of more than 900. Finding new *moai* gave us a great sense of adventure and discovery, but more critically these discoveries confirmed that the subtle features we had traced in the images were indeed roadways. In total, our ground survey documented more than twenty miles of ancient roads.[10]

Once they had been mapped over the island's landscape, we could look at their larger pattern. The roads emanate like spokes from the Rano Raraku statue quarry, extending across the island in at least four independent paths. One thing that is striking in contrast to roads in many parts of the world: these did not connect hubs of economic activity. Instead the roads began at the quarry and led to their many final destinations of *ahu,* most along the coasts. The roads were for statue transport, and probably not much else.

The evidence on the ground revealed that roads were not part of some overall planned network. Rather they are the remnants of paths that *moai* transporters took as they walked the statues across the landscape. In some cases, particularly where the terrain is steep, the paths converge. In large flat areas with few constraints for moving these giants, paths take idiosyncratic courses, sometimes winding over earlier *moai* roads, then straying from them. The frequency, intensity, and size of features comprising the roads also vary from place to place. We don't see a singular road form, but rather each path reflects the efforts of different people working over the centuries of making and moving *moai*.

The overlapping and independent pattern of *moai* paths suggests that the efforts involved in transport were not organized in an island-wide fashion that dictated use of a single path. Instead, moving statues may have been episodic, done over periods of time by a single group or different groups. Statue routes were not entirely predetermined by preceding paths and the groups involved appear to have been just as independent. Such a conclusion matches the pattern of statue styles seen at the quarry, reflecting multiple, contemporaneous teams of carvers following their somewhat independent traditions.

So it would seem that at least one part of the mystery of how the great statues were moved has been solved: they were transported along these well-constructed pathways. But transported how? As we've seen, the excavations of the roads offer numerous clues that some system using perhaps poles and ropes was used to move them along, but those are only the vaguest of notions. As we delved into this issue more deeply, we were able to connect a series of findings that offer a compelling—and truly remarkable—answer.

CHAPTER 5

The Statues That Walked

They walked, and some fell by the way.
—Katherine Routledge,
The Mystery of Easter Island,
1919

The statue quarry of Rano Raraku, carved into the cliffside of the volcano's crater, is an amazing sight. Hundreds of *moai* stand proudly along the crater's slopes, with many others at the base of the quarry that seem to be waiting their turn for transport. Yet others are buried a good way into the ground, some all the way up to their heads. Congregated here in so many states and positions, they conjure up a vision of scores of artisans at work and an eerie sense that the latter have just left, planning to return tomorrow. It is easy to imagine that their work ceased in a single moment, as if, as with Pompeii, some catastrophe occurred. We were to find that the state of preservation of the quarry's remains is a treasure trove of clues about the greatest mystery regarding the *moai*.

The walls of the crater are formed by a relatively easily carved tuff born of an explosive eruption that left behind compressed particles of ash and basalt stones. To do their carving, the islanders used crude hand axes known as *toki,* made out of basalt much harder than the quarry's volcanic tuff. Hundreds of *toki* litter the surface, and the quarry walls are covered in the markings made

Figure 5.1. The *moai* quarry at Rano Raraku.

from them during carving, each mark an evocative vestige of a single swing of an axe.

We know that statue carving often started with the face, as numerous faces peer out from the cliffside—nose, eyes, and mouth—in the process of being shaped. Carvers continued, completing the ears, chin, and neck, and moved on from there to the arms and the rest of the body. Many of the statues left in progress are standing vertically, but some were being carved horizontally. All together, with so many partial faces and bodies projecting from it, the face of the Rano Raraku cliff looks like an M. C. Escher drawing, with statues interlocked and overlapping in complex patterns.

The many partially completed statues tell us that once the front of a statue was finished, the carvers removed material from the sides and underneath, working from both sides and moving in toward a final ridge of tuff along the whole length of the emerging statue, which formed a keel that held it fast to the bedrock. For those statues carved high up on the cliff, it's likely that rope was then fastened around the statue from above. The final ridge of tuff was then cut away, and the statue was lowered down the slope to the crater's base. Large carved holes in the bedrock near the crater's summit are likely evidence of giant palm logs fixed into the cliffside as part of massive pulleys used to maneuver at least some of the *moai*. For the statues carved on the lower slopes, it seems they were slid down, leaving grooves worn into the crater's surface.

Not all statues successfully completed this journey. Some cracked along the way and were abandoned. We know that the head of one of these, still found at the quarry, was refashioned into a much smaller statue. Many other statues were left either lying or standing on the steep slopes. There is also an array of statues standing upright at the base of the quarry, many of them deeply buried so that just their heads are exposed. Often referred to as heads, these were a mystery, and continue to confuse many observers, but excavations first by Katherine Routledge and then by Thor Heyerdahl revealed that these were in fact full statues

Figure 5.2. *Moai* at the base of quarry in upright position with carving on back complete.

Figure 5.3. *Moai* at the base of quarry in upright position with incomplete back.

that ancient carvers had left standing in deep trenches once dug into the crater's lower slopes. The ancient trenches were either buried by the statue makers or filled in by sedimentation from centuries of erosion, leaving only the heads of the statues above the surface.

This suggests that by dragging and lowering the statues down the slope of the quarry, they were slid into the trenches, or sort of dropped gently into an upright position, where they could be erected easily to finish their carving. The outer edges of the quarry feature a series of these trenches, cut out perpendicularly from the slopes, filled with fragments and flakes of battered *toki*.

Figures 5.2 and 5.3 show statues located near each other on the bottom of the outer slope of the quarry. Each was carved from the upper slopes and lowered into trenches and left upright, but with the one on the left some additional carving had been done. The statue's head is thin and extended slightly forward. In addition, there are well-defined shoulders that are distinct from a somewhat "craned" neck. In contrast, the statue on the right has no neck or shoulders. Instead the long axis of the back is largely flat and continuous. This shape is a result of the long "keel" that once attached the statue to the bedrock from which it was shaped.

Once the statues were completed in terms of shape and balance, they were sent on a journey to their designated *ahu* along the roadways. As we have described, the *moai* roads are marked by a variety of constructed features—stone-lined curbs, leveled and flattened surfaces, cleared of stones—which must surely have played roles in moving the statues.

There has been much debate through the years about why so many statues were left at the quarry. Echoing Routledge and others, Jo Anne Van Tilburg claims that at least some were intended to remain at the quarry, never meant to be moved. In carefully studying all of the existing evidence, and making a series of our own observations and analyses, we found that the statues left at the quarry offer vital clues to answering the great outstanding question of just how the statues of Rapa Nui were moved.

Two basic notions about how they were transported have been

proposed. One is that they were moved upright by an intricate method of rocking them that made them "walk." The other is that they were placed horizontally, in the prone or supine position, on a wood platform of one kind or another and pushed or pulled in some fashion. A fascinating set of experiments has been conducted in attempts to test these competing ideas.

Thor Heyerdahl and members of his Norwegian archaeological expedition in 1955–56 began this process of experimenting with making and moving *moai*. Heyerdahl directed 180 islanders to place a medium-sized *moai* on a sled made of a forked tree and drag it a short distance using two parallel ropes (depicted in Heyerdahl's book *Aku-Aku*). This awkward "experiment" was met with polite skepticism by islanders who insisted, simply, that their ancestors had made the *moai* "walk." That the statues were moved vertically, with some apparently abandoned in their upright positions along the way, is also supported by some historical observations. For example, members of Captain James Cook's crew describe taking shelter in the shade of a standing statue (not on an *ahu* platform) near Ahu Oroi along the south coast. This statue later fell, as we find it today.

William Mulloy, the young American archaeologist on that expedition, would continue work on the island, and he took the walking notion to heart, proposing a method of swinging a semi-upright *moai* suspended by its neck with ropes from a wooden bipod, a contraption like a tripod but with only two legs. No one ever tested Mulloy's theory by experiment, and indeed it would have proven difficult to move the statues that way, to say the least.

More progress was made by Czech engineer Pavel Pavel. He had studied Heyerdahl's experiment, including the film of 180 islanders arduously dragging a *moai* a short distance, and he thought there must have been a better way. He envisioned walking *moai* by a method that at first seemed impossible. Imaginatively, he made clay *moai* models in miniature and discovered that their center of gravity often occurred at about one-third of their height, making them stable, like a bowling pin. His next step was to try moving a full-size standing *moai*. Using the grounds of the technical secondary school in the medieval town of Strakonice in then-communist Czechoslovakia, Pavel fashioned a concrete *moai* about fourteen feet tall and weighing a respectable twelve tons.

In July 1982 in Strakonice, Pavel and sixteen men working in two groups put his ideas to the test with this massive concrete *moai.* With ropes one group pulled the upright statue to tilt it on edge, while at the same moment the second group pulled to twist it into its first "step." By careful coordination, rhythmic pulls and twists wiggled the massive stone statue forward in a walking motion. For obvious reasons, this approach was quickly dubbed the "refrigerator method." Pavel's experiment showed that a small number of people, working in careful unison, could readily move a multi-ton *moai.*

In 1986, Thor Heyerdahl invited Pavel to join the Kon-Tiki Museum expedition to Easter Island to try out his theory on an authentic stone statue. Pavel's team prudently started off with a small *moai* that had been displaced in modern times. Abandoned behind the village post office in Hanga Roa, this lonely *moai* measured just over eight feet and weighed between four to five tons. A team of only eight people, carefully orchestrating pulls and twists of ropes, walked this *moai* forward with relative ease. But perhaps this small statue was too easy. The next challenge came with a larger *moai* measuring twelve feet tall and weighing about nine tons. The *moai* was padded with reeds and tethered with ropes, and just sixteen people jerked, tilted, twisted, and rocked this upright giant with remarkable success. It was a truly exciting moment for the Rapanui and Europeans alike. Heyerdahl later described the scene:

> Pavel spoke no language known to any of us, but he was a genius in making his orders understood by gesture, waving his arms and feet. As the experiment started it was difficult to tilt the statue over on one edge, but as soon as the workers began jerking rather than pulling steadily, the procedure became much easier. When the two groups, with more practice, succeeded in finding the exact moments of coordinating the sideways and forward jerks,

> they worked together rhythmically, easily and without strain. In this manner the experimental image wriggled forward as if it were "walking." . . . At first we were scared stiff that the men with the top ropes would pull so hard that the giant would capsize, but Pavel reassured us that the design of the *moai* was so ingenious that the colossus would have to tilt almost sixty degrees before it would fall over. . . . We all felt a chill down our backs when we saw the sight that must have been so familiar to the early ancestors of the people around us . . . an estimated ten tons "walking" like a dog on a leash.[1]

From these experiments Pavel estimated that small crews of experienced movers could transport *moai* as much as six hundred feet a day.

An elder Rapanui man witnessing the experiment, Leonardo Haoa Pakomio, explained to Heyerdahl that they not only had a song for walking *moai,* but a specific word to describe their unique motion: *neke-neke. Neke-neke* translates as inching forward by moving the body with disabled legs, or no legs at all. Leonardo demonstrated the meaning of *neke-neke* by alternately pivoting on the balls and heels of his feet, rocking slightly, and keeping his knees stiff. Heyerdahl responded: "what other language in the world would have a special word for walking without legs?"[2]

Not everyone was convinced, though. Ferren MacIntyre, a scientist with a breadth of interests at the National University of Ireland, took a critical look at the theories and pointed out that anyone who has moved a refrigerator can appreciate the merits of Pavel's experiments. But, he suggested, Pavel's method would damage the base of the *moai,* not to mention that moving them standing meant that "the line between meta-stability and disaster is uncomfortably thin."[3] To solve potential stability problems, MacIntyre proposed adding a curved timber rocking foot and side rig. Like Pavel, this method suggests the feasibility of moving *moai* in an upright position with relative ease and employing only a small team of workers. However, this method has yet to be tested in transport experiments.

Charlie Love, who did so much research uncovering the roads, thought that Pavel's method worked fine, but he believed it would prove impractical for long-distance transport.[4] Love reasoned that transport had to be accomplished with little damage to the finished statue, and in his own experiment he and his team used timber and ropes to construct a "pod" attached to the base of the statue, making it possible to roll it upright over several logs. The logs or "rails" could be brought up from the rear and placed in the path of the moving giant. Using a twelve-foot-tall, eight-ton concrete *moai* replica on the cold windswept plains in his home state of Wyoming, Love and his team rumbled the statue 150 feet along timber rollers in just two minutes. The dramatic experiment was caught on film and featured in a television documentary in 2000. Whether Love's experiment captured a method actually used remains unknown. But this successful experiment did reinforce Pavel's finding that a relatively small group, well coordinated to be sure, could move a giant standing *moai* significant distances.

In contrast, statue researcher Jo Anne Van Tilburg offered a different theory. In her view, horizontal transport of the *moai* was a matter of the logic of the statues' design. Van Tilburg started with a small one-tenth scale model of what she considered a "statistically average *moai*," dubbed "Sam" from its acronym.[5] Sam was scanned to create a digital image for computer simulations. In the same research, Van Tilburg used a digitized map to find optimal paths for moving Sam across the island from the quarry at Rano Raraku to Ahu Akivi.[6]

She then molded a twelve-foot, ten-ton statue replica in concrete on Rapa Nui to employ in her experiments. She and her collaborators used a wooden transport sledge constructed of beams of modern eucalyptus trunks arranged in a semitriangular ("A-frame") form. The team placed the statue in a horizontal position onto the frame with a lift from a modern crane, either faceup or facedown. Like Love's method, the rails could be repositioned from rear to front as the massive stone payload rolled forward. Misalignments and jamming of the log rollers in the first experiment led the team to formulate a modified design modeled after a Polynesian "canoe ladder." This altered sledge had "sliders" lashed to the triangular frame supporting the statue, keeping it from going astray. As it turned out, sliding over longitudinal rails worked, but rolling did not. Van Tilburg and her large team of islanders pulled the sledge hundreds of feet over level ground, impeded only by the limitations of manpower and rock outcroppings blocking the path of travel. She showed that about forty people could pull a ten-ton statue this way with little trouble.

Van Tilburg's experiment was featured in a 2000 *Nova* television documentary about moving and erecting *moai.* To provide a critique of the experiment *Nova* invited Vince Lee, an American architect who had studied how massive stone blocks could have been moved into place for Inca constructions like those at Machu Picchu in Peru. Lee recognized that once the statues reached their destinations, the steep stone-constructed *ahu* platforms perched on cliffs above the sea, the pullers, who had to be out front, had nowhere to go in order to pull the statue all the way up the steep incline leading to the *ahu.* So Van Tilburg's theory quickly met with significant skepticism. Lee hypothesized that using levers could solve the problem. He designed a set of ladderlike frames, where one frame sits atop another and levers operated between them provide the leverage for efficient movement of the statue up to the *ahu.* These levered sleds could be maneuvered by a small team, with the slider pieces being continually leapfrogged ahead, which would allow them to move *moai* over a variety of slopes and terrain. The levered sleds could "turn on a dime" and move a heavy payload up an incline onto the *ahu* without anyone working on the seaward side. Despite some problems with Lee's admittedly impromptu solution on Rapa Nui, his team of twelve men levered a six-ton monolith about fifteen feet in an hour and a half, each man easily moving one thousand pounds of rock.

About a year later, Lee perfected his experiment in Colorado, where, using the sled and slider ladders, twenty-five volunteers moved a thirteen-ton block of marble across an equipment yard, up a 25 percent slope, and rotated this payload 90 degrees, all in about two hours. The experiment convincingly replicated the maneuvering that seemed necessary to place multi-ton *moai* atop monumental *ahu* platforms just above sea cliffs. Lee admits there is no direct evidence that this method was actually used, but he points out that no other system he imagined could achieve these results.[7]

Experiments are just that. They show plausible solutions and sometimes reveal the possible and impossible. But we must be careful not to overinterpret the significance of *moai*-moving replications. In any case, making sense of experiments and building the best explanations of what happened must take the actual archaeological evidence into consideration. This is what we set about doing.

Some debate and speculation still surrounds why so many statues were left along the roads. Our close examinations of these particular *moai* provide an answer. We can identify a number of attributes showing that the statues along the roadways reflect abandonment due to failure that occurred during transport. One reason we know this is that none of these statues has completed eye sockets, and we know that carvers waited until the statues had arrived at their *ahu* destinations before adding the eyes. In excavations at Anakena, archaeologist Sergio Rapu discovered that one of the last steps of completing a statue was inserting white coral eyes with pupils of black obsidian insets. The statues were "blind" as they moved along the roads, until reaching their place of final enshrinement, where emplacing their eyes brought these stone ancestors to life. A second clue that the statues were abandoned due to failed transport is their position: nearly all are aligned with the road's direction and heading away from the quarry. As monuments placed along roads as markers or guardians that had fallen over time, one might expect these fallen giants to be found in random positions, with some lying across or diagonal to the road.

We decided to study these *moai* left by the roads more closely, and our surveying of the ancient roads helped us here. When we had followed the ancient paths, we had documented over fifty statues that lay either on or directly adjacent to the roads. As we now examined multiple attributes of those statues along the roads, we couldn't help but notice that many of them had broken into two or more pieces. In most cases, the breaks occurred where you would expect them: along the thinnest—that is, the weakest—portions of the *moai*. We also noticed that some statue fragments, especially the heads, were separated sometimes by a few feet or more, indicating that when the statue broke, the force propelled the fragments forward or back. Katherine Routledge observed the same. She wrote about the *moai* along the roads, "some single figures are lying unbroken," but "others . . . proved to be so shattered that no amount of normal disintegration or shifting of soil could account for their condition—they had obviously fallen."[8] The statues falling from a vertical position seemed the only way to explain this pattern in the fragments.

Furthermore, we noted, as Routledge had observed long before us, that some *moai* found along ancient roadways lay on their backs and some on their faces. If they were moved horizontally on sledges, as Heyerdahl first proposed and Van Tilburg continues to assert, then the positions on the roads should reflect how they

Figure 5.4. A view of a statue that has fallen during transport.

were placed on such contraptions. Moved horizontally, we should expect to find *moai* either faceup or facedown in more or less random associations regardless of their location, the slope of the road, or other features. But this is not what we found. Instead we discovered that their positions either faceup or facedown seemed to depend on whether they had been moving up an incline or down one.

When statues were heading upslope we usually found them resting on their backs, and when heading downslope they were on their faces.[9] A quick statistical test shows that this association cannot be explained as random. This is more support for the argument that the statues were moved upright. Rocking a standing statue back and forth would naturally result in falling forward on the downslope and backward on the upslope. Close study of the data gave us some intriguing additional support for the argument that the statues were moved upright. We made note of two statues on the southern part of the island that had come to rest at perpendicular angles, that is, somewhere between standing and fallen. These *moai* are partially buried, one at the base with his head found well above the ground, the other almost completely buried in a nearly vertical position. We conjectured that these *moai* must have fallen in transport and were then brought to their positions by attempts to re-erect them by excavating a pit. The logic is that a fallen statue could be re-erected by digging a pit near the base, sliding him into it to reestablish a vertical position, and then walking him out along a ramp of earth from the excavation. In the two cases we observed, the ancient project was apparently never completed and time has allowed the earth to fill in the pit around the *moai*.

We reasoned we might find further important evidence for upright "walking" by examining the wear-and-tear on the abandoned statues, and as we inspected them, another consistent pattern became clear. Guided by Sergio Rapu, we noticed that many *moai* had suffered damage along the sides of their base. This damage took the form of concave scars emanating from their base, what archaeologists call conchoidal fracture, *conchoidal* meaning cone-shaped. Conchoidal fracture produces flakes in stone. Physics tells us that conchoidal fracture results from substantial pressure that forces flakes from a stone. The fracturing we found at the edges of the *moai* bases, from the bottom up, was precisely the damage that would come from the statues being rocked back and forth in a vertical position.

Looking at the statues themselves, their locations, breakage, and positions on slopes, confirmed to us that they had walked from the quarry. The first experiments by Heyerdahl, and more recent attempts by Van Tilburg and Lee, showed ways to move massive loads horizontally with varying success, but these methods do not fit what we find in the archaeological record itself. This leaves us with the vertical method of walking proposed by Pavel, or perhaps those tried by Love and Lee, using pods, sleds, or devices with levers. But the theories for pods, sleds, ladders, or other wooden contraptions raise another problem. The only large tree known from Rapa Nui is the palm, *Jubaea chilensis,* or a close relative. Palm trunks have a thin, dense, brittle kind of bark and a soft, fibrous interior. Palms employed in the heavy work of *moai* moving would certainly see the bark crack, leaving the soft interior to be crushed. For the same reason, palms did not provide suitable wood for making canoes. So, not only were the *moai* moved vertically, we argue, but that was done without the aid of a wooden device, just as Pavel had demonstrated.

After learning that the *moai* were moved vertically, and without wooden contraptions, we wondered if the statues themselves might tell us more about their transport. On one of our visits to the statue quarry at Rano Raraku with Sergio Rapu, we began to find more answers. When we asked him about how the statues had been moved, Sergio pointed to the overall shape of the *moai.* Notice, he said, how the statues have a large belly and wide bases in the quarry, but they've slimmed down at the *ahu.* Also, note how their bases are angled, making them seem to lean forward slightly. These features, he commented, were not shaped for some kind of ancient aesthetic. Instead the *moai* were carved this way to move them. The problem was not just carving a *moai,* but carving one that could be moved. Reshaping that was done to the statues once they were erected at the *ahu* obliterated some of this telltale evidence. "*Moai* liposuction" prepared them for permanent display. Looking closely, we recognized that the *moai* standing on *ahu* had narrower bases and smaller bellies than those still on the roads, giving them a more imposing athletic physique. Excavations of *ahu* have revealed the debris of volcanic tuff shaved off *moai* in their cosmetic makeovers, archaeologically confirming the timing and locations of the changes. The *moai,* Sergio explained with pride for his ancestors, were engineered to move.

Engineered to move. We immediately realized this made absolute sense. Statue carvers could fashion forms of any shape, size, or configuration. Nothing but imagination constrained their artistic license, nothing unless the statue was to be transported without falling on its way to the *ahu.* Talking with Sergio and then contemplating *moai* along the roads, we realized that that is where the statue's center of mass (often referred to as "center of gravity") proved critical. The center of mass for an object represents a point where the mass is evenly balanced in all directions. In a regular object such as a sphere, this point is usually in the center, and if we could rotate an object around its center of mass, it would spin freely in all directions, as some globes do. A low center of mass is helpful for making objects stable in motion: skiers squat low to make tight, sharp turns and engineers design race cars to have a low center of mass to improve handling.

Thinking back to the many statues we had measured along the ancient roadways, we could envision how changing the center of mass by altering the shape of a statue would affect its ability to be moved. It was basic physics. Consider an object shaped like a cone, with its narrow end pointed down to the ground. The center of mass will be near the top. As long as the center of mass is directly over the point touching the ground, the cone remains upright, but it's a tricky balancing act. Just a little push and the cone will topple over.

If you've ever tried to balance a soda bottle upside down, you can readily imagine the scenario. This relatively high placement of the center of mass would be highly beneficial, though, if the goal were to move the object. A small input of energy would result in a big effect.

In contrast, when the center of mass is low, an object becomes relatively stable. Think of a bowling pin, for example. It takes a lot of movement, and energy, to tilt a cone with the wide side at the bottom and make it fall over, and the wider the base, the farther one has to tilt the cone.

To evaluate the properties of the center of mass of the *moai,* we both measured and photographed a number of them. This work resulted in three-dimensional wire-frame representations of the statues.[10] The figure below shows the result for a statue that had fallen during transport.

We measured the center of mass along each of three spatial dimensions. Obviously, the center of gravity was located in the middle of the statue in terms of its width. This position is not surprising, since the statues are generally left/right symmetric, and if the center of gravity were not in the middle, the statues would tend to lean to the left or right. The height of the center of mass is

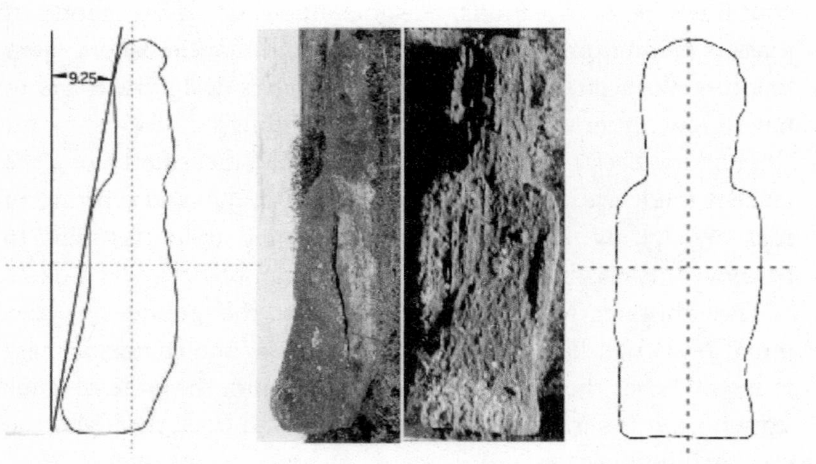

Figure 5.5. Profile and plan view of a statue found along a *moai* road. The dotted lines indicate the approximate location of the center of gravity.

also approximately in the center of the statue—midway between the base and the top of the head. But the center of mass in the depth dimension is remarkably forward relative to the base of the statue, just as Sergio had shown us. This was revelatory.

First, we cannot easily explain the peculiar location of the center of mass if the statues were transported horizontally. In fact, one would expect to find the center of mass toward the base, as this would facilitate raising the statue back up to its vertical orientation when placed on the *ahu*; it would effectively raise itself. There would also be no good reason for the center of mass to be located so close to the front.

On the other hand, this location of the center of mass makes a lot of sense for vertical transport through "walking." With the center of mass positioned at the front, rocking the statue back and forth is made relatively easy. This is similar to the physics of an inverse pendulum, which, unlike a regular pendulum, is turned upside down so the top swings back and forth from a fixed base. Both kinds of pendulums require very little energy to make them stay in motion for quite some time; they are very efficient at converting the energy we use to set them moving, and their own latent energy, called gravitational potential energy, into the kinetic energy of movement. As a pendulum swings, it travels an arc that spans from its highest point, when it has the most potential energy, through the bottom, when its velocity, and thus its kinetic energy, reaches a maximum, back to the level where it started when it very briefly slows to a stop. As the pendulum falls back down the other way, its potential energy is once again converted into kinetic energy, and the conversion is close to 100 percent.

The movement of an inverse pendulum also provides insight into the mechanics of our own walking. With each step you transform yourself into an inverted pendulum. When you pick up your leg to walk forward, you pivot on the foot that is placed on the ground. As you pivot, your center of mass—somewhere in the belly—follows the path of an arc. Your forward foot eventually hits the ground and your arc slows to a stop in that direction. At that point your kinetic energy is at a minimum—but your potential energy is at a maximum. As you fall forward into the next step, the stored potential energy is converted back into kinetic energy, and you accelerate again. This is the basic physics of walking. Moving large upright objects such as refrigerators and *moai* takes advantage of the same principle.

The forward and midlevel positioning of the center of gravity in the *moai* allowed them to be easily tipped back without falling over. The farther back a statue can be tipped, the farther the center of gravity can be swung forward. This maximizes the amount of potential energy as the statue is rocked and thus the farther forward it can be moved on each next swing. Of course, the center of gravity cannot be placed too far forward or the statue will easily fall facedown. With a margin of error provided, the position of the center of gravity almost—but not quite—to the forward edge is optimal for statue walking, but accidents were clearly to be expected. Once the statues were walking, the problem may have been stopping them.

Movement in this way also accounts for the shape of the statue bases, at least for the larger *moai*. The figure on page 91 shows the base of a statue that fell while being transported. The statue is lying facedown; its large belly is visible along the bottom edge. The shape of the base is related to how the statues were moved.

The front edge (toward the bottom of the photo) is strongly rounded, as that would be the portion of the base that would rotate while the statue was tilted. The back of the base is slightly rounded. The sides of the base, on the other hand, are relatively straight. This shape provided a long edge for initiating the tilting of the statue. The large surface area minimized the potential failure of the material caused by the weight of the statue. The curved front edge, however, reduced surface area and thus the friction as the statue rotated. Less friction meant easier moving, and less wear to the bottom of the *moai*.

Not all statues are shaped this way. Some have much wider bases, and others are more symmetric. Smaller statues, in particular, are quite variable. The taller and larger *moai*, however, are

Figure 5.6. A view of the base of a *moai* fallen in transport along an ancient road.

more similar in shape and form to each other. And this makes perfect sense, since the larger the statue, the more important it would have been that the shape conform to the constraints dictated by transport. While a small statue might be moved regardless of its shape, the larger statues would have been a very different story. To those most skeptical we concede that it is possible that while the larger statues consistently have a form enabling vertical transport, they were nonetheless moved horizontally. However, we conclude that such a consistent center of mass that would have aided "walking" makes their horizontal transport most unlikely. Of course, the precise details of how the ancients moved the statues are still being teased out of the details of the archaeological record. But we believe it's clear that the archaeological facts, like the island folklore, tell us the *moai* "walked." They moved upright, traveling slowly and steadily over challenging terrain, propelled by teams of probably only fifteen to twenty people. Taken alone, this fact may seem inconsequential. But it is a critical piece of the larger puzzle for the true story of what unfolded on the island. The notion that making and moving a *moai* must have engaged hundreds or thousands of workers amounts to pure speculation. There was no need for a large population to support the making and moving of these monuments.

Walking the *moai* would have required cooperation, but not a powerful paramount chief overseeing a complex organization of conscripted carvers and pullers. The expediently constructed *moai* roads and the statues found along them reflect, we believe, the work of small-scale social groups. And walking the *moai* did not require vast amounts of timber for wooden sleds, rollers, or sliders. It was not a reckless mania for *moai* that exhausted the island's forest and tipped the ecological scales toward catastrophe. Solving the mystery of how the ancient islanders moved the *moai* pointed us to a dramatically different story for Rapa Nui, one that was written in stone and awaiting our best efforts to decipher it.

CHAPTER 6

A Peaceable Island

The wolf also shall dwell with the lamb, and the leopard shall lie down with the kid; and the calf and the young lion and the fatling together.

—Isaiah 11:6

Many of the native populations of the Pacific have a long and well-documented history of aggression and fighting. The upper valleys of Fiji are famous for hilltop fortresses, Tahitians constructed massive war canoes for interisland raids, and Hawaiian chiefdoms battled each other in long-standing wars involving hundreds of warriors with fearsome battle clubs. The evidence of fighting in the archaeological record is clear: weapons, defensive structures, and skeletal evidence indicative of violent deaths.

It is not terribly surprising that violence and competition are common in many island populations. Given the limited resources on most islands, and the natural tendency of populations to grow over time, it seems that competition would be just about inevitable. Sooner or later, individual needs will conflict over the availability or access to one or another limited resource, whether it be land, food, mates, or raw materials. History and common sense dictate that scarcity inevitably leads to conflict.

It would therefore be logical to conclude that the archaeological record of Rapa Nui would exemplify conflict. The small size of the island meant that the population would have quickly filled the landscape, and the 1,500-mile ocean gap between the island and its closest neighbor—tiny Ducie Island—meant that before long there was no way that the growing population could expand further. With just sixty-four square miles of surface area on the island—some of which wasn't suitable for cultivation, such as the steep slopes of the volcanoes—it would not have taken long for available land for cultivation to become a premium. When we first arrived on Rapa Nui, we expected that the archaeological record would divulge plenty of evidence of conflict, but it didn't. Instead our archaeological investigations have shown that Rapa Nui's history is notable for its lack of violence.

Of course, there is no way of directly viewing violence in the archaeological record; we must make interpretations from whatever artifacts we can find. Typically, the remains from human conflict fall into three classes. The first and the most direct form of evidence are skeletal remains, and in many places around the world we find many of these: bodies with no heads, arrowheads in skulls, smashed skulls, and so on. But the skeletal remains of prehistoric Rapanui show few signs of lethal trauma. In a study conducted by physical anthropologists[1] of a collection of nearly five hundred individuals from the island, only a very small fraction showed evidence of any trauma and almost all of those revealed evidence of healing that occurred after the event. Just 2.5 percent had evidence of injuries and most of these were restricted to the skull. Males showed about twice the trauma of females, and these marks were consistent with the use of clubs and incisions by sharp blades. In most instances, though, the evidence suggests the attacks did not lead directly to the deaths of the individuals. A few skulls had obsidian pieces still embedded in the bone, yet the individuals clearly survived the injury, as there was healing of the bone around the wound. Given that at least some of these remains date to the historic period in which violence by Europeans is well documented, the evidence from skeletal remains for prehistoric violence is actually even less substantial.

The second kind of evidence of violence we look for is remains of weapons. Given the physical properties required to pierce bodies and damage internal organs, archaeologists can identify objects that were designed specifically for killing, though where game animals exist, weapons for hunting would have much the same appearance as those built for conflict with human enemies. So we also consider where such artifacts are found, and their associations, as indications that they were used in violence between individuals.

On Rapa Nui, it is often assumed that obsidian tools known as *mata'a* are evidence of lethal weapons. These are roughly triangular implements that have a stem chipped on one end for hafting to some kind of shaft. How lethal *mata'a* would have been, however, is open to debate. While the edges are sharp, their shape is not appropriate for piercing, and we should be cautious before concluding that they were used in warfare.

Consider the traditional technology of Rapa Nui. Unlike other locations in the prehistoric world, the only regularly shaped objects that are thought to be potential weapons on the island are the *mata'a*. Working with her Rapanui informants, Katherine Routledge described *mata'a* formed by "obsidian [that] was knapped till it had a cutting edge, and also a tongue [the stem], which was later fitted to a handle or a stick. The various shapes assumed were dignified by names, fourteen of which were given, such as 'tail of a fish,' 'backbone of a rat,' 'leaf of a banana.'"[2]

The obsidian used to make *mata'a* formed in the vents of volcanoes during eruptions when lava reached the surface and cooled quickly. This process produces a glassy material that is brittle but has exceedingly sharp edges. When available, obsidian was used worldwide for the creation of high-quality cutting tools. In fact, obsidian blades have been measured to be 210 to 1,050 times sharper than a steel scalpel.[3]

There are at least five discrete areas on Rapa Nui where obsidian can be readily obtained. These sources are located in the southwest corner of the island from ancient vents of the volcano known as Rano Kau. Two of the sources are on the exterior slopes of the primary volcanic crater island, one is on an adjacent cinder cone, and two are on tiny islets, called *motu,* that are just offshore. Overall, obsidian was one of the few raw materials that would have been relatively abundant and probably freely available on Rapa Nui.

Mata'a are produced from pieces of obsidian ("flakes") that were knocked off larger chunks ("cores") using a dense basalt stone as a hammer. Archaeologists call this technique "hard hammer percussion." The flakes are modified on one side through additional percussion to produce a narrow, semirounded stem between one and two inches in length. Although we have no direct archaeological evidence, it is presumed that this stem was used to haft the tool to a wooden shaft that would serve as a handle. One line of evidence we have that suggests that the *mata'a* were mounted to a handle is the presence of dulled edges along the stem. These dull edges would be necessary to prevent the stem from cutting through cord used in hafting.

The edges opposite the stem are largely unmodified though sharp from the initial stages of production. The overall shape of these edges is variable. Rather than being commonly pointed, *mata'a* are wildly inconsistent and vary from rounded to subangular to angular to complex. The stem is what *mata'a* have in common, since the "working" edge is irregular in shape.

Today *mata'a* are found in large numbers across the surface of Rapa Nui, comprising one of the most common shaped tools (as opposed to flakes or the products of manufacturing) that we find during our field surveys of the island. The Father Sebastian Englert Anthropological Museum on the island houses literally thousands of these artifacts collected over the years, and yet we routinely encounter hundreds more on our field surveys.

European visitors have commented on the presence of objects like *mata'a* since the late eighteenth century. In 1770, for example, the Spanish explorer Don Francisco Antonio de Agüera y Infanzon, of the Spanish frigate *Santa Rosalia,* noted that the native populations had no arms, but had cutting instruments of sharp-edged stones that might serve as weapons. Similarly, Captain Cook reported that on Rapa Nui individuals "had lances or spears

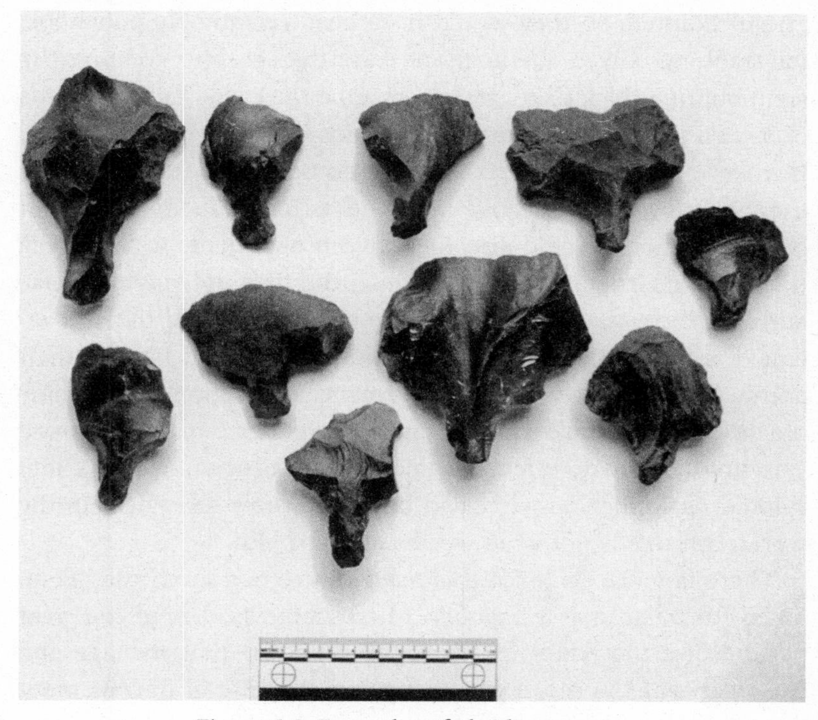

Figure 6.1. Examples of obsidian *mata'a*.

made of thin ill-shaped sticks, and pointed with a sharp triangular piece of black glassy lava."[4]

Beginning with these early European interpretations, *mata'a* are still thought to be spear points or some kind of implements used in combat.[5] Assuming for the moment that this interpretation is true,[6] their shape and overall design would be important if used in combat. And if *mata'a* were used in personal combat, it would have proven quite risky. *Mata'a* are simply inappropriately shaped for use as projectiles. With asymmetric shapes and heavy points relative to shaft length and thickness, these objects would be inaccurate and lousy as thrown projectiles. In addition, as stabbing tools they are too thick and only haphazardly (and rarely) pointed, so they would have been remarkably poor tools for stabbing. Given the skill and care that go into creating the stem out of a thick flake, we can assume the crude business ends of *mata'a* were not the results of a lack of talent or knowledge. If it were necessary, the prehistoric occupants of Rapa Nui had the means to make effective implements for weapons. Indeed, if competitions were held directly between opponents wielding the thick, blunt, irregular-shaped *mata'a,* the first individual to create a thin, symmetric pointed blade would have had the equivalent of a "weapon of mass destruction" and would have quickly dominated the island. Given performance differences, the archaeological record would show a rapid convergence in form toward variants of a single effective type of *mata'a*—thin, pointed, and symmetric—much as we see in projectile tools elsewhere in the world. But this is not what we see on Rapa Nui.

There is more evidence that *mata'a* were not specifically fashioned for or used as weaponry. Microscopic studies of the wear patterns on the edges of *mata'a* have shown that damage and use-wear was the result of cutting and scraping of fibrous plant materials.[7] These studies point to a conclusion that *mata'a* were predominately used for cutting and scraping, kinds of activity related to the use of plants, mostly, and not to spear points. In fact, this evidence finds support in observations made by Cornelis Bouman, Dutch captain of one of Jacob Roggeveen's ships, the *Thienhoven.* Bouman remarks that the islanders did not recognize the function of the Dutch knives they were shown and yet cut "bananas with a sharp black stone."[8]

We have to evaluate, however, the possibility that while *mata'a* were not stabbing weapons, they could still have been deadly, and other lethal implements were available to the islanders as well. Implements used as weapons could have included sharp sides of *mata'a* to slash and cut, stone adzes, wooden clubs, or simply stones, and these might have been lost to the record or might not offer clear signs of their use as such. Items made out of wood, for example, may not have been preserved, and stones don't generally show signs of being used as weapons. We have good reason to believe that they were, though, as a variety of firsthand accounts by early European visitors describe incidents of stones being thrown by islanders.[9]

We also have early descriptions of *mata'a,* or objects like them, thought to have been used specifically for people cutting and slashing at each other. For example, in 1770, Agüera y Infanzon noted that "we never saw their bravery put to the test, but I suspect they are faint-hearted; they possess no arms, and although in some we observed sundry wounds on the body, which we thought to have been inflicted by cutting instruments of iron or steel, we found that they proceeded from stones, which are their only [weapons of] defense and offense, and as most of these are sharp edged they produce the injury referred to."[10] We will return to the implications of nonlethal violence in the next chapter, but here it is important to note that if *mata'a* served as weapons at all, the observation made at the time of the Spanish voyage was of obsidian tools used in a fashion consistent with our studies—cutting and slicing, not stabbing.

The last kind of evidence for conflict that archaeologists look for are remains of defensive structures. In Polynesia defensive features are most commonly found in the form of walls, earthworks, and fortresses. The small island of Rapa Iti in the Austral Islands, for example, has ample evidence of prehistoric warfare visible in elaborate hilltop fortifications.[11] On Rapa Nui, however, there is no evidence of hilltop strongholds or any other kind of fortification that would serve as defense. There is a formation that has been described as the Poike Ditch, often interpreted as a defensive feature. But geologists have shown that the "ditch" was produced by the confluence of the lavas flows of the Poike and Terevaka volcanoes.[12]

The caves of the island are often described as having been modified for defense, though it is unlikely they could have served this purpose when opponents were so aware of the details of the island's landscape. As shelter from Europeans, they may well have been effective hiding places, but not against other islanders.

Thus we are left with a dilemma: how could the prehistoric

Rapanui have avoided deadly conflict over resources? We believe the answer to this question is relatively simple: it did not pay to escalate conflict to levels that resulted in lethal violence. Sometimes, as it turns out, the best long-term strategy in a situation such as that on Rapa Nui is for people to show restraint.

REASONS NOT TO KILL

Much has been made about the propensity of humans to kill each other. Violence and murder are startingly common in contemporary human societies. Sir Arthur Conan Doyle, author of the Sherlock Holmes stories, remarked that "strange indeed is human nature. Here were these men, to whom murder was familiar, who again and again had struck down the father of the family, some man against whom they had no personal feeling, without one thought of compunction or of compassion for his weeping wife or helpless children, and yet the tender or pathetic in music could move them to tears."[13] James A. Froude, an English historian and writer, argued that "wild animals never kill for sport. Man is the only one to whom the torture and death of his fellow creatures is amusing in itself."[14] Similarly, Mark Twain declared that "man is the only animal that deals in that atrocity of atrocities, War."[15]

But some brilliant work by evolutionary biologists and social scientists has shown that in certain circumstances, humans often choose to be doves rather than hawks. British evolutionary biologists John Maynard Smith and George Price were fascinated by the manner of competition observed in a number of animal species, which can be intense but most often stops short of lethal violence.[16] Take the case of mule deer. When competing over mates, the mule deer stags often engage in displays of aggression. The fighting is preceded by a series of escalating challenges in which first the stags "roar" at each other. Then, if one does not back down, they challenge one another by walking in parallel, often exposing vulnerable areas in a display of toughness. Finally, they may turn, lock antlers, and engage in pushing contests. But only rarely do these competitions lead to injury and death. Similar kinds of restraint have been observed in many other species, including snakes that wrestle yet withhold the use of their fangs.

Maynard Smith and Price wondered why this restraint in violence would have evolved. If aggressive individuals and their heirs benefit in these contests by eliminating competition and gaining mates, then the aggressive types should eventually overtake the weaker, more passive ones. If these animals defeated passive individuals in every contest, then sooner or later only aggressive types should remain.

Investigating this puzzle, they worked out that being the strongest is not always the best strategy, specifically when there are potential costs involved in competition. Maynard Smith and Price imagined a hypothetical model with just two types of members of society, "Hawks" and "Doves." Hawks always escalate every conflict and never back down. Doves, on the other hand, always back down. They then imagined how each type would fare as they encountered one another in a contest. Two Doves that meet each other make a showy display, but one of them ultimately retreats. While one wins the contest, neither sustains injury. When Doves and Hawks meet, Doves back down and Hawks win. Finally, when two Hawks encounter one another, escalation occurs until one or the other is injured or killed.

What does this have to do with the comparative lack of violence on Rapa Nui? Maynard Smith and Price modeled interactions between Hawks and Doves in a formal way that has come to be known as "game theory," in which they kept track of the results in a series of interactions among and between the Hawks and Doves.

To see how this plays out, imagine that you are a Dove in a group of mixed Doves and Hawks. There is competition over some resource (for example, land, raw materials, food, mates) with an arbitrary value of 10. This value simply symbolizes some measure of "gain." Also imagine that if a fight occurs there is going to be a cost to the loser. So, if one is a loser in a contest between Hawks, then not only does one get nothing (winner take all), but there is a penalty from injury, damage, or loss of resource. Here let us say that the cost is arbitrarily 30. In these contests, we will say that all the Hawks are of equal quality so the odds in any contest are fifty-fifty whether one or the other individual will win. Obviously, this value of the cost means that gains are made with substantial risk: the winner gets 10, while the loser gets negative 30.

In this scenario, as you regularly interact with other individuals to acquire resources, sometimes you encounter another Dove and you resolve the issue peacefully by sharing the resource. At other times you encounter a Hawk and you yield to his aggressiveness. Each time you have such an interaction you record what you gain, if anything. Over time your resources accumulate from the gains made with your interactions with Doves. Nearby is your friend, a Hawk known to always escalate conflict in any encounter. This Hawk also has a series of encounters with other individuals and gets gains. At a gathering you and the Hawk compare notes as to how well you are doing. You report that over the last nine encounters, you fared like this:

Table 6.1. A Series of Encounters Between You and Some Neighbors

You . . .	**Encounter . . .**	**and Receive**
Dove	Hawk	0
Dove	Dove	5
Dove	Hawk	0
Dove	Dove	5
Dove	Dove	5
Dove	Dove	5
Dove	Hawk	0
Dove	Hawk	0
Dove	Hawk	0
	Total	**20**

The sum of the resources that you gain for that set of encounters is 20. Your friend the Hawk notes that he interacted with the same set of individuals. His encounters, however, went like this:

Table 6.2. Summary of Results After Your Set of Encounters

Your friend . . .	**Encounter . . .**		**and Receive**
Hawk	Hawk	(Friend wins the contest)	10
Hawk	Dove		10
Hawk	Hawk	(Friend loses the contest)	–30
Hawk	Dove		10
Hawk	Dove		10
Hawk	Dove		10
Hawk	Hawk	(Friend wins the contest)	10
Hawk	Hawk	(Friend loses the contest)	–30
Hawk	Hawk	(Friend loses the contest)	–30
		Total	**–30**

Your friend the Hawk is not happy. Though he is the more aggressive one, his position is worse than yours. Your friend's total resource gain over the series of interactions is −30. He is worse off than when he started! This simple exercise reveals that although aggressive Hawks always win in individual contests with Doves, over the long run the cost of competing with other Hawks results in doing worse than Doves.

There are a couple of points to be made based on this observation. First, cost matters. The degree to which being a Dove provides an advantage over Hawks is largely a function of the magnitude of the cost of losing. If there is little or no cost involved, then Hawks will do better than Doves. This means that aggressive individuals reap rewards whenever the gains outweigh the costs of losing. In social scenarios, for example, bullies benefit if there is no teacher around to punish bullying. We would always expect escalation to dominate in situations where a loss has no consequence. On the flip side, if costs are high, nonescalation should prevail and Doves will dominate.

Now, we can complicate the model a little bit. Rather than "be" a Dove or a Hawk, let individuals be free to choose one or the other strategy. They can choose based on what they believe will be successful at any point in time. When one has a choice about which strategy to play, it turns out that it always pays best to be the opposite of whatever one thinks one's opponent is going to be. If confronted with a Hawk, be a Dove. When confronting a Dove, be a Hawk.

The problem with a strategy like this is that in most cases there is no way in advance to guarantee knowledge about what one will confront. Could this be a Hawk who will back down when confronted? Might this apparent Dove be an individual who only looks meek, but ultimately will turn out to be a Hawk? How can one be sure? But given the scenario played out above, we can figure out what choice one should make over the long run: be a Dove.

RAPA NUI DOVES

It is reasonable to ask at this point: are these "thought exercises" really a way to explain why Rapa Nui was apparently not driven by violence? Are these simplistic models capable of informing us about the real world? We believe that in the case of Rapa Nui, these insights from evolutionary theory are particularly powerful given the limited resources, the lack of opportunity for the islanders to leave the island, and what would have been the high costs of interpersonal violence.

There were many factors of life on the island that we argue constrained violence due to the costs that would have been borne over the long run by engaging in it. From the technology of weaponry available to the islanders, to the degree of interrelatedness among the island's population, to the physical environment, we see good reasons for the islanders to have refrained from violence.

With only stones to throw, wooden clubs, spears, or *mata'a*, each weapon available to the islanders required them to confront their opponent in a close-up, direct, and personal way. Such face-to-face conflict would entail substantial risk of serious injury or death. Consequently, when we evaluate the Dove and Hawk scenario for Rapa Nui, we should expect to see a significantly decreased proportion of Hawks over time.

The islanders were also strongly incentivized to consider the longer-term consequences of attacking anyone. For one thing, there would have been no secrets on the island; Rapa Nui is so small that word of any misdeed would have traveled all over the island fast. This is still true today, as we know from our many seasons of doing fieldwork on the island. Someone noticed just about everything we did, and almost as soon as we did it, the news made it across the island. If one of our students had a little too much fun at one of the late-night bars, by morning we were already being told about the details even before the student woke up.

In such an environment, one can imagine the futility of attacking a competitor or attempting to lie, cheat, or steal. Relatives would have sought retaliation and there would have been no means of escaping the consequences.

Also, because the island was colonized only once, we know that most of the population was closely related. As a result, any conflict between any two individuals would have had social implications at some level through family relations and traditions and could have sparked escalating costs, since the results would have affected not only the two involved, but their families and longstanding arrangements. Communication moves rapidly in small communities and even the simplest slight could trigger effects that ripple across families along patterns of relatedness. Even today we see these patterns of conservative social interaction among the native population of Rapa Nui. In fact, given that the population fell to as low as just 111 individuals in 1877, the present Rapanui descendants are related in a pattern that likely resembles that of prehistoric times.

Despite the small size of the community, the genealogical and political landscape of the contemporary population is exceedingly complex. One dimension of this complexity is that, in one form or another, everyone seems to be related to everyone else. Factions are not obviously split across kin groups and one finds that for any particular issue—whether political office, land, economic development projects, indigenous rights, or archaeological conservation—family members can be found on all sides. This kind of tight-knit community is common on small islands, as we have seen time and again living in villages and doing archaeological fieldwork in remote parts of Samoa, Fiji, and New Guinea. Such complex social relations are particularly pronounced on Rapa Nui.

What this social situation prompts is cautious decision making and explicit efforts to avoid conflict. Being passive (a Dove strategy) is the norm, particularly since there is almost nothing you can do that does not affect someone to whom you are related. Thus we see the development of strong social norms seeking consensus, rather than resorting to violence. Indeed, early European visitors noted the tendency for islanders to avoid conflict. In 1770 Agüera y Infanzon, for example, took notice of the way islanders did not fight over personal property. He stated, "they are so fond of taking other people's property that what one man obtains another will take from him, and he yields it without feeling aggrieved: the most he will do is resist a little, then he loosens his hold of it and they remain friends."[17] These kinds of passive interactions are consistent with a social landscape that seeks to avoid conflict and favors compromise over confrontations. On the island there is little to be gained from conflict, and everything to be gained from cooperation.

Even the landscape of Rapa Nui was not conducive to strategies of violent behavior. The island is so small and so barren that there would literally have been no place to hide, at least not for long. Though the island's caves later offered protection against slave traders, as said before, as a refuge from other islanders they would have offered a temporary haven at best, since their locations would have been widely known.

Considering all of these issues, the lack of warfare, or even much less lethal violence, in prehistoric Rapa Nui culture can be demystified. Though the conjectures of horribly escalating violence between groups on the island have made for a good story, the archaeological record simply doesn't bear out this scenario, and the lessons of evolutionary theory and of anthropology also argue against it. But how and why, then, did the population of the island collapse? And if violence didn't escalate on the island over competition between groups, then why did the islanders spend so much of their time and devote so much of their energy to the building of their monumental statues? Exactly what role did the statues play in the life of the island? It is now time to explore the answers to this long-standing mystery.

CHAPTER 7

Ahu and Houses

> Who were the clients on this hard-bitten islet that could afford to draw the bulk of the talent and muscle of the population away from the life-and-death struggle for food? Who were the patrons that could afford to copy nature and bury their dead in such huge and magnificently architected platforms? Who could pay for such finely-cut and fitted cyclopean masonry, as only the great prehistoric empires of Greece and Crete, or that of Egypt, or that of the Incas at the zenith of their power have indulged in?
>
> —John Macmillan Brown,
> *The Riddle of the Pacific,* 1924

One of the most remarkable aspects of the archaeology of Rapa Nui is certainly the apparent paradox we see in the spectacular cultural achievements compared to the island's diminutive size, remote location, and paucity of resources. Part of the reason we find Rapa Nui prehistory so improbable lies in common preconceived notions that a place lacking resources could hardly support elaborate community projects in art, architecture, or religion. Part of Western heritage is the idea that such feats of cultural elaboration—particularly aspects of culture that are not related to reproduction, food production, or other kinds of obvious basic human needs—are the product of surplus wealth and increased social "complexity." In other words, those populations more evolved are those with the greatest control over their resources. This is the concept of cultural evolution, a notion distinct from Darwin's idea of biological evolution. The idea of topping the evolutionary ladder has its roots in nineteenth-century Europe, and Europeans believed themselves at the high end of the ladder. At the core of the concept of cultural evolution is the premise that the achievements of other populations are meaningfully measured relative to those of Europe.[1]

Rapa Nui stands in marked contrast to the expectations that would follow from this notion. The island has obvious cultural elaboration, but with no evidence of vast agricultural production and food surpluses, massive populations, hierarchical governments, or the other trappings of the Western notion of "highly evolved" society. We can try to explain this anomalous occurrence in two ways. First, we might posit that at one point the necessary ingredients for cultural elaboration on this scale were present on the island. This is the tack taken by Thor Heyerdahl, who was convinced that Incan colonists from South America were the makers of the *ahu* and statues. His assumption goes even further, also claiming that the Incans responsible for the cultural florescence on Rapa Nui were ultimately the descendants of colonists with European origins who taught Native Americans the secrets of "advanced culture."[2] For Heyerdahl, simply tracing the "cause" of Rapa Nui culture back to Europe solved the apparent paradox of cultural achievement. Leaving its racist assumptions aside, empirical support for this argument is entirely lacking.

Many others, of course, have reasoned that there must have been "advanced" aspects to Rapa Nui at one time but that they vanished in a great cataclysm. Early-twentieth-century travel writer John Macmillan Brown assumed that the island must have once been larger and that vast agricultural fields, essential to support the thousands who made the *moai,* must have sunk beneath the sea.[3] But the reasoning in such arguments is circular: cultural elaboration occurs in "complex" societies with abundant resources; thus if we see cultural elaboration, the society must have been "complex" and wealthy. Lack of evidence for wealth and "complexity" simply points to some past event that wiped them out.

We can do better. Before exploring the flaws in this reasoning further, however, we should first take a closer look at the issue of whether the islanders would have needed outside help to erect their monuments, and also examine the best evidence for the role the statues played in the culture and whether they should be considered evidence of a "complex" or highly centralized society, as has so often been conjectured. Doing so will reveal that the techniques the islanders used did not rely on outside expertise, as Heyerdahl argued, and that the society was not in fact centralized in its social and political organization.

PLATFORMS FOR STONE GIANTS: *AHU*

We have considered how the *moai* were made and transported, but we haven't looked closely yet at the question of how the *ahu* were constructed, and since this was pivotal to Heyerdahl's assertion of Incan involvement, we should turn to this now. William J. Thomson, the paymaster of the USS *Mohican,* conducted the first systematic inventory of *ahu.* In 1886 Thomson walked the coastline of Rapa Nui and identified 113 *ahu.* With assistance from ranch manager Alexander Salmon and local hired guides, Thomson recorded names for each of the platforms as identified by the islanders who accompanied him. These names, like many of the place-names on the island, come from tradition and local knowledge. However, we cannot be sure of how far back the names were used since they were only written down so late. By 1886 the population of the island consisted of only 155 individuals, of which only 111 were adults (68 men, 43 women), and given the loss of population, we have to expect there was some loss of inherited knowledge. Thomson's descriptions of the *ahu* included measurements as well as comments on their state of preservation, and his inventory preserved many of the names for the *ahu,* which are still used, names such as Ahu Tongariki and Ahu Tahai. The list has been expanded over the years, and the most recent inventory was conducted by Norwegian archaeologist Helene Martinsson-Wallin in 1994. Her work is the most complete inventory to date, cataloguing 150 *ahu.*[4]

The building of the *ahu* was most inventive, requiring engineering solutions that would allow the platforms to support the weight of the giant *moai.* Construction began by assembling massive boulders that formed the interior foundations of the platform. Once the interior of the *ahu* was largely complete, the boulder foundations comprising the interior were filled in with rubble. Construction of the seawall followed. This is an immense wall facing seaward and forming the "back" of the platform relative to the statues facing forward. The seawall provides the height to the structure and is crucial to the support for the weight of the statues. The style of construction of these walls varies, ranging from large boulders stacked in an irregular fashion with smaller blocks infilling gaps, to cut basalt blocks that tightly fit together on their outer edges. But while there is this variability, there are a number of elements common to all *ahu.* First, the seawalls lean inward slightly at an angle of about five degrees from vertical. This angle distributes

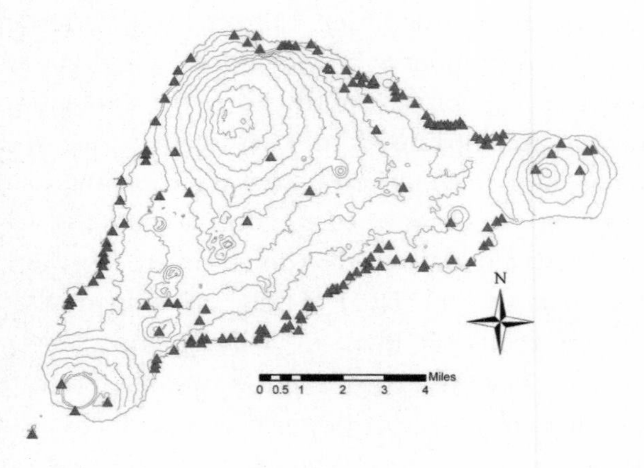

Figure 7.1. The locations of "image" *ahu* based on studies by Helene Martinsson-Wallin.

the weight of the statues downward and allows for a more stable outer wall. Second, the strength of the wall depends on the friction between the blocks that hold it together, and the more friction the better. Consequently, *ahu* are constructed with large basalt blocks that fit snugly against each other. Often blocks were cut and smaller, shaped rocks wedged in between to increase friction.

Ahu walls occasionally were made of extremely well-fit stones, placed very tightly together. Ahu Vinapu, located in the southwest corner of the island, is the best existing example of this precisely fit wall construction. Here the massive seawall blocks fit together so well that gaps are only millimeters in width. We know that other examples of this kind of wall once existed on the island, and an *ahu* known as Ahu Hanga Roa, which was somewhere in the vicinity of the modern town, was described by James Cook in 1774. He writes that *ahu* walls "are built, or rather faced, with hewn stones of a very large size; and the workmanship is not inferior to the best plain piece of masonry we have in England. They use no sort of cement; yet the joints are exceedingly close, and the

Figure 7.2. Seawall at Ahu Vinapu. This wall is often cited as evidence of South American influence on Rapa Nui prehistory.

Figure 7.3. *Ahu* seawall on north coast of Rapa Nui.

Figure 7.4. Seawall construction of an *ahu* platform
(Ahu Tepeu, west coast).

stones mortised and tenanted into one another, in a very artful manner. The sidewalls are not perpendicular, but inclining a little inwards, in the same manner that breastworks [fortifications], etc. are built in Europe."[5]

This observation highlights why Heyerdahl was inclined to argue that the quality of these walls is linked to the influence of South Americans who, he believed, had first colonized the island. Indeed, the regular and fitted cut stones have at least some similarity to the massive masonry walls made by the prehistoric Inca. Found in abundance in ancient cities such as Cuzco, Ollantaytambo, and Machu Picchu, Incan masonry is characterized by finely cut granite blocks with regular sides that fit together like a puzzle. The similarity to *ahu* walls is undeniable, especially when one limits comparisons to sites such as the exterior part of Ahu Vinapu.

The resemblance, however, turns out to be only superficial. Close examination of the details of wall construction shows that careful alignment of the edges of the blocks is limited to just the front surface of the wall. In places where the inside of the *ahu* walls are exposed, we can see that the sides of the seawall blocks are irregularly shaped and the fine fit is little more than a facade. Stones are shaped in place so that the front edges match up. The sides of the blocks often angle inward, and the space between the blocks is filled with rubble. This practice is in distinct contrast to Inca stone walls, where all sides are fully and regularly dressed. These details of stone block construction in South America suggest that specialized stonemasons shaped blocks before they were placed in walls. In contrast, on Rapa Nui blocks were placed and shaped during a single process of wall construction. Each stone was roughly dressed, then aligned with adjacent blocks that were modified, and finally fit together so that the front edges met. Holes or gaps were patched with small rocks that were often themselves shaped. One gets the impression that wall construction on Rapa Nui was done with relatively small groups of individuals working on single areas, rather than a massive industrial layout consistent with the complex specialization of the Inca.

From the top of the platform, *ahu* typically slope downward to a

Figure 7.5. A reconstructed Ahu Nau Nau with red scoria lintels adorning the upper platform and an array of *poro* stones on the front slope.

lower ledge. This slope is often decorated with a carefully arranged array of *poro* stone (dense, rounded basalt boulders found in the tidal surge zone of the island). According to Sergio Rapu, when he excavated one *ahu,* Ahu Nau Nau at Anakena Beach, the *poro* stones were colored white, likely from crushed coral that was prehistorically applied to their surfaces.[6] This white color, however, was washed off or faded after being exposed to the elements. Given that the areas between the stones consists of a layer of red scoria gravel, the original front side of the *ahu* was likely quite colorful.

Though we do not know the details of activities performed at *ahu,* there is evidence both from the archaeological record and historical accounts that they were ritual sites. The area in front of the *ahu* is often called the plaza, typically an expansive area leveled out of the undulating and rocky terrain. In many cases the material necessary to construct the plaza greatly exceeded the material that formed the platform portion of the *ahu,* in total requiring a high degree of engineering and labor.

There is little in the way of reliable information about the kinds

Figure 7.6. Ahu Tongariki showing *moai, ahu* platform, and plaza in the foreground.

of rituals performed at *ahu*. The fact is that our archaeological knowledge is limited and eyewitness accounts provide only a tiny glimpse of how these features were incorporated into Rapa Nui life. Roggeveen had probably the best opportunity to make observations, but his account states quite little. He writes: "What form of worship of these people comprises we were not able to gather any full knowledge of, owing to the shortness of our stay among them; we noticed only that they kindle fire in front of certain remarkably tall stone figures they set up; and, after squatting on their heels with heads bowed, they bring the palms of their hands together and alternatively raise and lower them."[7]

We have a couple of other minor bits of information that suggest attitudes about the monuments and events staged at them, but we are largely left to guess about their meaning to the islanders. During Felipe González de Haedo's 1770 visit, Sublieutenant Don Juan Hervé, first pilot of the *San Lorenzo,* wrote: "Throughout the island, but especially near the seabeach, there are certain huge blocks of stone in the form of the human figure. They are some twelve yards in height, and I think they are their idols. They could not bear to see us smoke cigars: they begged us extinguish them and they did so, I asked one of them the reason, and he made signs that the smoke went upwards; but I do not know what this meant or what he wished to say."

Don Francisco Antonio de Agüera y Infanzon, chief pilot of the *Santa Rosalia,* described similar sensitivity to the *ahu* and statues: "The sculptured statues are called Moay by the natives, who appear to hold them in great veneration, and are displeased when we approach to examine them closely."[8] From these and later descriptions, it seems clear that *moai* represented deified ancestors of the Rapanui, probably at the scale of clans who held territories over parts of the island. Descriptions collected from native informants by Katherine Routledge suggest *moai* may have been related to ceremonies involving some kind of constructed figure made of perishable materials. Known as *paina,* these perishable figures might have been temporary versions of the more permanent *moai.*

Agüera y Infanzon was the first to describe these stick and grass figures, in 1770:

> They have another effigy or idol, clothed and portable, which is about four yards in length: it is properly speaking, the figure of a Judas studded with straw or dried grass. It has arms and legs and the head has coarsely figured eyes, nostrils, and mouth: it is adorned with a black fringe of hair made of rushes, which hangs half-way down the back. On certain days they carry this idol to the place where they gather together, and judging by the demonstration some of them made, we understand it to be the one dedicated to enjoyment, and they name it Copeca.[9]

Fleuriot de Langle of the La Pérouse expedition reported a similar feature constructed near one of the *ahu*.

> We found near the last of [the *ahu*] a kind of layman or effigy of reeds, representing a man ten feet high, and covered with a white manufacture of the country; the head of a natural size, the body thin and the legs pretty exactly proportionate, and a net hanging to its neck in the shape of a basket covered in white clothes, and apparently containing grass. By the side of this sack was the figure of a child two feet long, with the arms crossed and legs hanging down. This layman, which could not have stood there many years, was perhaps the model from which statues are now erecting to the chief of their country.[10]

Later informants told Katherine Routledge that these reed figures were placed in the circular depressions located at *ahu*.[11] Of course, it is difficult to ascertain whether this was always the case or if the raising of reed figures was the only activity that took place near the circular depressions. Obviously, however, the *ahu* and the *moai* were the focus of much attention, given the labor invested in their construction and maintenance.

A CENTRALIZED SOCIETY?

That the islanders invested in building statues is not surprising, and we do not think their number can be taken as evidence that the island once had a highly centralized social organization.

We've noted previously that statues and platform construction are not uncommon across eastern Polynesia. Examples of large stone statues are known, for example, from the prehistoric record of Pitcairn Island, though unfortunately, the European settlers of the island destroyed those they found.[12] British naval officer Frederick William Beechey visited Pitcairn in 1825 and reported that the first Europeans to land on the island, the *Bounty* mutineers, saw "three or four rudely sculptured images, which stood upon the eminence overlooking the bay."[13] French merchant and diplomat Jacques Antoine Moerenhout traveled to Pitcairn in 1837. He described multiple statues eight to ten feet tall and found on stone platforms. When Routledge visited Pitcairn in 1915,[14] after her research on Rapa Nui, she found only one fragment of a statue torso, which had been pushed off a cliff onto the beach below. While part of an expedition with Alfred Métraux to Rapa Nui, Belgian archaeologist Henri Lavachery recorded a single statue, hidden under the plank floor of an islander's house.[15]

Large stone statues are also known on the islands of Nuku Hiva and Hiva Oa in the Marquesas and on Raivavae in the Australs. In addition, large constructed platforms and courtyards (often referred to as *marae* or *heiau* in the Hawaiian Islands) are well-known throughout many of the islands of eastern Polynesia, including the Societies, the Cooks, and the Tuamotu atolls. Monument construction and the forms of cultural elaboration seen on Rapa Nui, therefore, are widespread in Polynesia. Indeed, the similarity of forms points to a strong shared ancestry among these people. The platforms tend to be rectangular in shape, and the statues typically have large heads and arms that wrap around the belly.

Consequently, as we've discussed before, the puzzle about Rapa

Nui's statues is not their mere presence, but rather the magnitude of the investment the islanders made in them. For such a remote and resource-poor place, an immense amount of labor was put into shaping and moving large rocks. Given that populations who colonized East Polynesia had shared ideas and knowledge when they reached new islands, what about Rapa Nui and its population can account for its spectacular amount of statue building?

This is where the argument has come in that the society must have evolved into one with a strong central authority that could compel the islanders to devote so much of their time and energy to statue building. But cultural elaboration, ranging from mound construction, to monumental architecture, and other forms of labor investment not directly related to food production, is found throughout prehistory and among populations with many kinds of organization and subsistence. The famous megalithic structures of Stonehenge were associated with small-scale, dispersed communities.[16] The same can be said of the many ancient megalithic structures found across Europe and India. Surveying the global evidence, there does not appear to be a single set of conditions necessary for the emergence of cultural elaboration. In fact, when we again examine the actual archaeological record of Rapa Nui, it shows that rather than a society clustered in a number of relatively large settlements, such as villages, the population of the island was instead widely dispersed.

One common assumption made about sedentary societies, whether prehistoric or contemporary, is that people live in organized villages, towns, and cities. This form of settlement is known as "nucleated," in which a community is clustered together within a single local setting. These kinds of settlements tend to have relatively high population density and produce large, dense archaeological sites. But while nucleated settlement patterns are commonly seen today, they are not the only way in which sedentary communities can be configured. Communities can also be widely dispersed, with families being economically self-sufficient, except that they share some kind of communal activity.

A famous case of this kind of settlement patterning can be found in prehistoric eastern North America. Dotting the landscape of wooded valleys of the Ohio and Illinois rivers, archaeologists have found evidence of an extraordinary cultural phenomenon that took place between 1100 BC and AD 400. During this time populations known as Hopewell in Ohio and Adena in Illinois constructed a remarkably large number of geometrically shaped earthworks—huge circles, long straight lines, massive conical mounds, and other geometric shapes—that stretch up to 1,700 feet across. Excavations of conical-shaped mounds have uncovered log-lined tombs that were often filled with exotic materials from areas far away from Ohio, including obsidian from Wyoming, marine shell from the Gulf of Mexico, and copper from the upper parts of Michigan. The amount of investment required to fill these tombs is mind-boggling: individuals walking and canoeing from Wyoming to Ohio moved at least several tons of obsidian.

This is why archaeologists expected that for Hopewell and Adena they would find evidence for hierarchical organization to account for the coordinated efforts needed to make these monuments. Immediately, however, they were puzzled: none of the areas near the mounds and earthworks appear to have been inhabited. These ceremonial features are entirely devoid of domestic material. So where did the people live and how did the labor and material involved in these "empty centers" get there? It took some careful analysis to determine that the landscape around the mounds was dotted with small hamlets—clusters of just a few houses—marked by low densities of artifacts easily overlooked when one is looking instead for large nucleated settlements. Thus the Hopewell and Adena communities lived in dispersed settlements and people gathered seasonally at mounds and earthworks to share food and perform ceremonial activities. Despite the cooperation and degree of investment in labor, however, they did not have a centralized or controlling economic or social organization. Rather it seems that participation in the communities provided sufficient benefits through resource sharing and information to allow the system to be successful and persist.

Some kind of centralized and shared activity is a key feature of dispersed communities. This activity brings communities together to share resources, redistribute materials, or exchange information. The nature of the centralized activity reflected in the archaeological record varies, including everything from mound construction to elaborate burials, ceremonial activity, and monument building such as on Rapa Nui. The communal activity serves to structure interaction and merges separate domestic units into a single integrated community, but only in these particular ways.

One advantage that dispersed communities have over nucleated ones is that they can potentially exploit a larger and more diverse resource area. In cases where productivity is unpredictable over a region given rainfall, local soil conditions, or limitations of plant growth, dispersed communities tend to be more stable since they can average sometimes uneven returns, across the entire community.

RAPA NUI SETTLEMENT PATTERNS

Our field surveys conducted on the south-central and northwest coasts, and earlier work by Patrick McCoy in the southern and southwestern areas of the island, show that islanders in prehistoric Rapa Nui lived in this dispersed pattern of settlement.[17] There were no centralized villages or towns. Instead these detailed archaeological surveys reveal repetitive sets of stone structures and other features reflecting the day-to-day life of ancient times. These often include *manavai,* vast rock "mulch" planting areas, chicken houses (*hare moa*), earth ovens (*umu*), and house foundations and areas paved with smooth, water-worn boulders (*poro*). Families lived in huts made of grass and a thin wooden superstructure. An excellent description of these houses is provided by Jacob Roggeveen, who visited one of these houses under construction in 1722.

> Their houses or huts are without any ornamentation, and have a length of fifty feet and a width of fifteen; the height being nine

feet, as it appeared by guess. The construction of their walls, as we saw in the framework of a new building, is begun with stakes which are stuck into the ground and secured straight upright, across which other long strips of wood which I may call laths are lashed, to the height of four or five, thus completing the framework of the building. Then the interstices, which are all of oblong shape, are closed up and covered over with a sort of rush or long grass, which they put on very thickly, layer upon layer, and fasten on the inner side with lashings (the which they know how to make from a certain field product called Piet, very neatly and skillfully, and is in no way inferior to our own thin cord); so that they are always as well shut in against wind and rain as those who live beneath thatched roof, in Holland.

These dwellings have no more than one entrance way, which is so low that they pass in creeping on their knees, being round above, as a vault or archway; the roof is also of the same form. All the chattels we saw before us (for these long huts admit no daylight except through the one entrance-way, and are destitute of windows and closely shut in all round) were mats spread on the floor, and a large flint stone which many of them use for a pillow. Furthermore they had round about their dwellings certain big blocks of hewn stone, three or four feet in breadth, and fitted together in a singularly neat and even manner; and, according to our judgment, these serve them for a stoop on which to sit and chat during the cool of the evening.[18]

One remarkable and common kind of house on the island has a long elliptical shape with large carved basalt stones forming the foundation. These structures are known as "boat-shaped houses," called *hare paenga* or *hare vaka. Paenga* refers to long rectangular stones shaped from basalt that measure from two to six feet in length, six inches to a foot across, and one to three feet high. They typically have a series of "cupules" carved into the top surface. These cupules are relatively shallow, usually measuring a couple of inches in depth. The *paenga* are embedded into the ground to form a long oval shape similar to an outline of a canoe. Two

U-shaped *paenga* stones mark the ends of the outline. Overall, *hare paenga* range from about fifteen feet in length to massive versions that stretch well over one hundred feet.

The entrances of the *hare paenga* are usually found in the center of one side. Outside the entrance, many *hare paenga* have semi-oval–shaped "patios" made of embedded *poro* stones. These houses are also often found in clusters, with the doorways generally facing toward communal *ahu* sites. Some researchers have claimed that these structures were for "elite" individuals and separate from houses for commoners.[19] This assertion, however, is unsupported by the archaeological record as well as ethnohistoric accounts.

Our excavations of the deposits within *hare paenga* foundations located near Ahu Akahanga and along the northwest coast of the island revealed few, if any, artifacts associated with domestic activity. Early historic visitors to the island made observations consistent with our findings, remarking that all they saw inside these houses were some gourds. Thus it is likely that *hare paenga* were not primary habitations but may have had a special, perhaps

Figure 7.7. Stone foundation of a typical "boat-shaped house," or *hare paenga*.

ritual, role on Rapa Nui. Given their size and narrow shape, they would not be well suited for much other than sleeping. Their general clustering around *ahu* suggests that they are somehow related to *ahu* activities, though exactly how is unknown.

This distribution of domestic features reflects dispersed settlement patterns. The farther inland one goes, the lower the density of domestic archaeological features, such as rock mulch areas and *manavai*. As we travel inland from the coast, we also find an increasing array of artifacts that indicate specialized economic activities as opposed to domestic life: basalt quarrying, carved bedrock water basins, small rock alignments with terraces, and so on.

Variations in the style of artifacts and architectural remains suggest that groups shared ideas locally, and that communities were relatively small subsets of people compared to the island population as a whole. For example, open-air earth ovens (*umu*) are common features throughout the island. *Umu* were the primary means of cooking food. While *umu* are functionally identical and formed by a pit dug into the ground enclosed by rocks, their construction varies in shape. The surface stones can be arranged in rectangular, pentagonal, circular, and irregular shapes. None of these shapes affects the performance of the oven; each is simply the result of shared (or inherited) cultural preferences, that is, a local style for doing things.

As it turns out, on Rapa Nui, "local" can be very local. When we plot the distribution of different *umu* shapes, or styles, we find that their numbers vary from place to place. Some areas, for example, have an abundance of rectangular ovens, while other areas have more circular ones. We find similar variations in styles in the details of *ahu* construction, *pukao, moai,* and even in *mata'a* tool forms.[20] For example, while *ahu* share many features, the details of seawall construction, presence of red scoria lintels, and other specific items vary from platform to platform. Our analyses of these *ahu* features have shown that they vary little by little as one moves along the coastline. *Ahu* in close proximity will be more similar than those farther apart. *Moai* and *pukao* also appear to vary over distance, although some variation also reflects changes in styles

Figure 7.8. An earth oven, or *umu*. The circular arrangement of stones marks the top of a pit in which food was once cooked.

through time. It is remarkable that styles of artifacts and architecture could vary from place to place on such a small island.

We find the same kinds of "microgeographic" variations evident in studies of the human skeletons from the island. George Gill, a physical anthropologist at the University of Wyoming, has found skeletal anomalies largely restricted to specific areas of the island.[21] For example, skeletons buried on the north coast at Ahu Nau Nau commonly had a kneecap (the patella) that was shaped in a bipartite fashion. On the south coast, on the other hand, Gill found many individuals who had a fused sacroiliac joint—a condition in which the triangular bone that sits below our lumbar vertebrae (the sacrum) joins with the top of the pelvis (the ilium). And on the west coast, Gill found a high incidence of distinctive shapes for the supracondylar foramen of the humerus—an opening in the upper arm bone. These kinds of physical traits are passed through families and suggest that, at least to some degree, marriage practices were localized on the island.

Finally, a recent study of ancient human genetics undertaken by John Dudgeon of Idaho State University as part of his dissertation work at the University of Hawaii has found the same kind of patterns.[22] Dudgeon was able to extract ancient DNA from 222 individual teeth from skeletons recovered from thirteen locations around the island. Like Gill, Dudgeon found patterns of genetic similarity indicating that interaction was highly localized, especially for males, but that there was some movement of females between areas along the coast.

The Rapanui population did not stay entirely within the bounds of local communities, however. A study of a large sample of crania by Vincent Stefan, a former student of George Gill, showed that marriages were not just limited to local areas, but must have included some island-wide interaction.[23] In particular, as in Dudgeon's finding, women appeared to move between areas more than men did. Overall, however, archaeology plus human biology shows that smaller, largely self-sufficient and self-contained communities occupied relatively small territories across the island.

Given the dispersed pattern of settlement, we can conjecture that similar to the building of North America's mounds, the construction and activities surrounding *ahu* and *moai* played a key role in integrating smaller, local communities into a larger single one. Found spaced in a repetitive way, *ahu* occur at a larger scale than the domestic units scattered across the landscape. Their regular distribution suggests their importance for social groups, perhaps at the level of tribes or clans, and certainly for groups smaller than the entire island population. These ceremonial structures and their statues were focal points for communities. Investing in the communal activities that probably included labor for *ahu* construction, *moai* transport, as well as other *ahu*-based ceremonies such as cremation or burial of the deceased, integrated the community through sharing that would benefit those who participated. Understanding the archaeology on the ground and the social interactions it reflects, it becomes clear that no paramount center of centralized political control ever existed on ancient Rapa Nui.

One caveat for describing and explaining the archaeological record of Rapa Nui is that today we see artifacts and structures that accumulated over the past eight hundred years of the island's human history. While it's tempting to think that provides a snapshot of the way things were, we must bear in mind that a complex history of events unfolded over centuries. Knowing precisely which structures were built and used at the same time remains a difficult challenge. And we know that elements of settlement, architecture, and economic activity changed over time. The archaeological record of the island is a palimpsest of remains from multiple overlapping communities that existed over time and across space. Careful dating and analysis will be required to tease out more details about the settlements as they evolved through time. But the evidence is already clear that the portrayal of an island dominated by a strong central ruling authority, or torn asunder by warring tribes, is simply wrong.

So, if the islanders were not compelled to engage in excessive statue building by some strong central authority that ruled them, why did they do so much of it? Why would they have invested so greatly when that would seem to have taken too much of their time and energy away from cultivating crops? In the next chapter we will show that there are a number of good reasons to believe that rather than being an overwhelming burden on the sustainability of the island's culture, the building of so many statues was integral to the islanders' having been able to sustain their culture so well for so long.

CHAPTER 8

The Benefits of Making *Moai*

The present natives take little interest in the remains. The statues are to them facts of every-day life in much the same way as stones or banana-trees. "Have you no *moai* . . . in England?" was asked by one boy, in a tone in which surprise was slightly mingled with contempt; to ask for the history of the great works is as successful as to try to get an old woman selling bootlaces at Westminster the story of Cromwell or of the frock-coated worthies in Parliament Square. The information given in reply to questions is generally wildly mythical, and any real knowledge crops up only indirectly.

—Katherine Routledge,
The Mystery of Easter Island, 1919

Moving multi-ton statues necessarily required the cooperation of groups of individuals. And this cooperation had to be sustained over weeks, months, or years. Given that the labor invested in *moai* resulted in no obvious immediate survival benefit to people who were living at the subsistence level, the question of why they agreed to participate in this activity is a pressing one.

We are often asked why the people on the island "made those giant stone heads,"[1] and our answers are often disappointing since we avoid speculating about what the islanders were thinking when they made them. Any account that relates to "what people were thinking" would simply be a story, arbitrary and distorted by biases of our own times. But this doesn't mean that we can't provide any answers about functions the statues may have provided to their makers.

Of course, part of the answer—as we have discussed previously—can be found in Polynesian traditions that colonists carried when they arrived on the island. Constructing stone platforms and statues was part of shared cultural traditions. But as we've said, those on Rapa Nui engaged in the activity to an extraordinary degree. One answer for why they did so comes from research in evolutionary biology of what is called *costly signaling*.

Costly signaling is simply a kind of communication. It conveys honest information about qualities being advertised that benefit both signalers and recipients. The kinds of qualities advertised might include control of resources ("wealth"), health, competitive advantage, or other attributes. The "costly" part refers to the idea that the signals impose a cost that would be difficult to fake, thus ensuring accuracy. The benefit for signalers comes with opportunities for attracting mates or allies as well as communicating abilities in cooperative or competitive interactions. We could think of signaling as actions to gain status and the benefits that accompany it. For recipients of signals, the payoff comes from the ability to evaluate the qualities advertised, and make informed decisions about mating, cooperation, competition, and so on. Costly signaling benefits both sides, since the cost means the signals usually reflect truth in advertising. Costly signals often take the form of visual displays.

Ahu and *moai* certainly fall into the category of costly signals broadcast in visual displays. The fact that *moai* and *ahu* are such obvious and powerful visual displays suggests that these features provide benefits to the individuals who invest their energy in their construction. It would be difficult to fake the time, energy, and intelligence required to make and move gigantic statues. So to answer the question of "why statues?" we need to examine the conditions under which investment in displays might be favored. What is being signaled with the display and to whom is it being displayed?

Signals can significantly mitigate the amount of outright conflict people engage in. As we've seen, biologists, studying the dynamics of such competition, have found that in a conflict over resources or mates, two rivals both benefit if the likely loser walks away from a physical contest, since both conserve energy and avoid injury. Thus there is a mutual advantage if a threat accurately conveys information such as the ability, resolve, and strength necessary to win a physical contest. Many organisms, including humans, present a particular "posture" when challenged by an aggressor. For human males, this posture is characterized by standing tall, with chest out and shoulders back, appearing prepared to fight.

In general, when one adopts a posture in the face of a threat, one increases the risk of a rival's attack. A hostile posture signals that one is prepared to meet an aggressor's challenge. While false signals might be tempting, the risk involved is high. Consequently, the display of "toughness" in the face of an aggressor is effective only if the individuals who do it are confident that they will win. If the signal can be tested, then there is little reward for cheating, creating a good correlation between an individual's signal of willingness and his ability to win a physical contest. Overall, such signaling can reduce the risks of competition on both sides and for this reason it is common in the natural world.

The study of signaling in human populations has a long history. In 1899, sociologist Thorstein Veblen first described human acquisition and display of resources in terms of the signals they send to potential competitors and mates.[2] In his book *The Theory of the Leisure Class,* Veblen describes the acquisition of fancy goods that do little more than waste money and resources as "conspicuous consumption." For example, while regular utensils are perfectly adequate for eating food, many people invest in utensils made of silver or gold. Obviously, there is no added functional value for these items other than that they display one's ability to have them. This kind of wasteful extravagance, he argued, allows people to demonstrate high status. Of course, less wealthy individuals often attempt to attain greater status through the acquisition of these signals or by finding cheaper replicas.[3]

Many types of our behavior can potentially be explained through costly signaling. For example, driving a Lamborghini, having a dinner party, and philanthropy all command attention and confer prestige because they provide reliable evidence of an individual's control or accumulation of resources. It is the direct cost of these activities that guarantees that they are reliable signals of access to resources. Reliability of the signal must be testable so that others can gauge their own relative status and decide whom to consider a potential mate, ally, or enemy. Thus, despite conflicting interests, there are benefits to the sender and the receiver of these prestige signals: receivers learn the abilities about individuals around them, and transmitters gain prestige and status.

THE PAYOFFS OF BET-HEDGING

In addition to the effectiveness of the *moai* as signals, we think that they served another vital function on the island, in helping to limit the size of the population.

Generally we would say that in the survival-of-the-fittest competition of evolution, the most successful are those who produce the most heirs in the long run. But that doesn't mean the most successful are those who have the most offspring in any given generation. Having the greatest number of children possible may not be the best reproductive strategy, particularly in an environment of scarcity, since the more children there are at any given time, the more strain is put on limited food resources. In environments where resources are limited or unpredictable, the most successful will tend to be those who have limited the number of their offspring, a strategy biologists refer to as bet-hedging. Many animal species engage in this reproductive strategy. Research on reproductive behavior among many animals and humans has shown that the trade-off for smaller numbers of offspring is the ability to provide greater parental investment. In these situations, life expectancy increases and more young survive to reproduce.

What is the link to statue making? Cultural elaboration is

one mechanism that can produce the benefits of bet-hedging. Any form of cultural elaboration requires an investment of time and energy in activities that do not have a clear or direct role in enhancing survival or enabling reproduction. The classic example of burial ceremonialism provides a case in point. Archaeologically, the investment in burials for the dead has its origins well into the late Pleistocene, perhaps as early as fifty thousand years ago. At this time we see the first evidence of individuals being buried in structures such as tombs and mounds with rare and exotic material as grave goods. All of the energy and labor involved in preparation of a burial site and the acquisition of goods to place within a burial are effectively "wasted" because living individuals, and their future offspring, can no longer use them. While burying one's dead has important functional aspects such as sanitation, preventing disease, and keeping predators at bay, the additional investments of grave goods have little direct relation to survival or reproduction.

The vital link between these kinds of cultural elaboration and bet-hedging strategies is that they represent large expenditures of time and energy on the part of the participants. We see archaeological manifestations of bet-hedging with cultural elaboration in many locations around the world. As we mentioned earlier, the tradition of building massive earthworks and mounds with elaborate burials of the Adena and Hopewell in eastern North America provides one example. The Hopewell and Adena populations faced many of the same environmental constraints as those on Rapa Nui. They were sedentary populations living in a relatively unpredictable environment, with a dependence on wild and cultivated plants for subsistence. These American populations responded in much the same way as the Rapanui. Combining dispersed settlement systems with investment in ceremonialism, they traded short-term productivity for long-term stability. In the American and Rapa Nui cases, the forms of cultural elaboration are highly visible and serve as clear displays on the landscape. In fact, much of what we tend to think of as the remains of bet-hedging strategies around the world involves highly visible mounds and monuments—from megalithic constructions across Europe and India, to the massive prehistoric mounds of Japan, Africa, and North America.

Under conditions where resources are consistent over time, there will be a disadvantage to high investment in cultural elaboration relative to those who focus their resources on reproduction. In such cases we expect to see less cultural elaboration within a population through time. But, in contrast, under conditions where resources vary, bet-hedging strategies will be more successful than other variants, and thus we expect to see increases in the amount of cultural elaboration.

In the short term, this "wasteful" strategy seems to be self-defeating, because the population doing it would appear to be sacrificing the benefits of reproductive success, leaving them at a competitive disadvantage. But ultimately evolution is a numbers game. In the end, those who produce the most grandchildren are most successful relative to everyone else. While it seems counter-intuitive, nature shows us that where we find unpredictability in resources, those who have fewer offspring in any one generation tend to have more descendants overall.

In making this point, we have to be careful to emphasize that we are not saying the islanders consciously decided to spend more time making and moving statues in efforts to take time and resources away from reproduction. The role of the statues as costly signals probably explains why so much time was devoted to them. But the benefits of bet-hedging offer a strong explanation for why spending so much time and energy on the statues didn't threaten Rapanui survival, and indeed probably helped to sustain it. In considering what leads to evolutionary success, those reaping the benefits of any given practice need not understand why it contributes to their survival.

People often engage in customs that happen to have evolutionary benefits, but without ever thinking about them in that way. Take the case of the Naskapi Native Americans who live in the northern reaches of the Labrador Peninsula on the east coast of North America. Traditionally, they were hunters who tracked herds of caribou for their subsistence. These populations live in an area that is quite marginal for most kinds of human habitation. Their existence, at least historically, strongly depended on hunting caribou; little else in this harsh environment could support human communities.

Anthropologists who studied these people in the 1930s described a curious practice that was central to their hunting strategies. To plan their hunting trips, Naskapi people would consult animal bones. The shoulder blade of a caribou—the scapula—was considered the most effective of all the bones for determining hunting strategies. Once the scapula was scraped of meat and hung to dry, it would be suspended over hot coals. As the bone heated, it cracked and burned into endlessly variable sets of patterns and markings. By tracing the patterns, hunters would predict where caribou herds would be and would thus ensure a successful hunt. While we might dismiss this as just superstition, the practice is serious since the stakes are high: failure to predict the location of caribou herds could easily mean starvation for families in this remote and barren environment.

Predicting the future by reading the cracks on a heated scapula is known as scapulimancy, and it is known also in Asia and other parts of the world. In Korea, for example, pig and deer scapula were used for divining future events as early as 300 BC. In China, the use of scapulimancy dates to at least 1500 BC and may be related to the origins of the Chinese writing system.

Clearly, since logic tells us there is no known relationship between the patterns of cracks and future events, it is possible that scapulimancy persists simply as a harmless cultural tradition. Such good-luck charms are common in most cultures: lucky rabbit feet, religious medallions, lucky hats, "rally monkeys," the number seven, and so on. In the case of the Naskapi hunters of the Labrador Peninsula, however, there is more going on than meets the eye.

In the 1950s, an anthropologist named Omar Khayyam Moore described Naskapi scapulimancy not as a simple cultural idiosyncrasy, but as an essential mechanism related to the long-term suc-

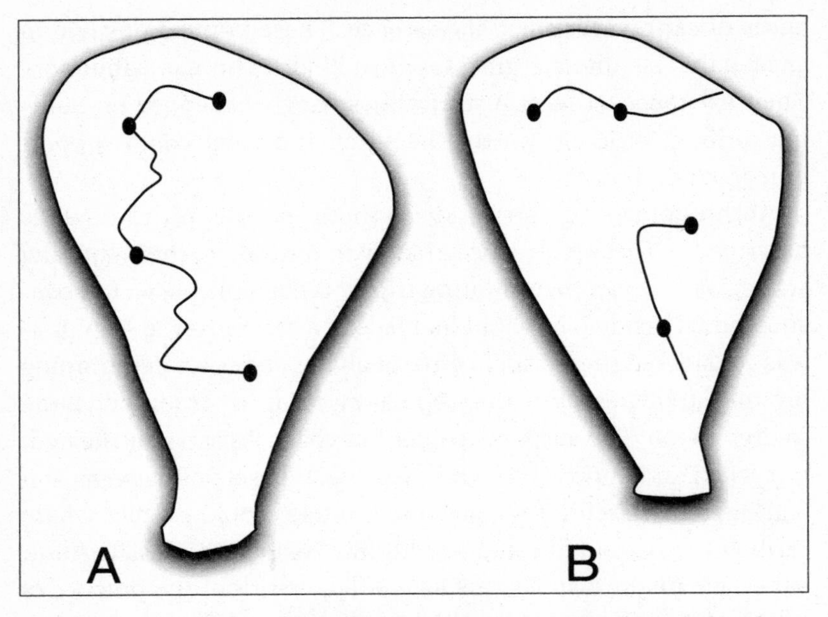

Figure 8.1. Caribou scapulimancy. Scapula are cleaned and placed over a fire. The patterns of burning and cracks provide a divination of future hunting conditions. On the left (A), the pattern of cracks and spots shows the location of caribou and successful hunts. On the right (B), the patterns of cracks show an unsuccessful hunt since the hunter's trail does not join that of the caribou.

cess of Naskapi populations.[4] The problem, Moore explains, is not necessarily in predicting the specific spot where caribou can be found, but rather in making sure that enough variability is introduced into hunting plans so that success comes neither too rarely nor too often. In the long run, adding variability or an element of randomness is the key to success. Imagine a scenario in which a hunter had a run of bad luck. Each day this hunter might venture out thinking he knows where the caribou herds will be found based on previous experience. Repeatedly taking that route, however, can quickly lead to the animals' being sensitized to the hunter's strategy, leading the hunt to eventually fail.

Adding some randomness into the plan—different routes taken at different times—prevents the development of anticipatory responses. Likewise, repeated failures will be minimized as the randomization provides a way of searching the landscape without following the same patterned strategies, consciously or unconsciously. A bit of randomization prevents a hunter from making the same bad choices over and over. Paradoxically, it also prevents hunters from too much success, which could result in overhunting.

Scapulimancy provides just this feature and does so just as we use random number tables today to minimize inherent biases in choices we make. Moore pointed out that "it is difficult for human beings to avoid patterning their behavior in a regular way. Without the aid of a table of random numbers or some other randomizing instrument, it is very unlikely that a human being or group would be able to make random choices even if an attempt were made to do so."[5]

Consequently, scapulimancy "works" and persisted in the Naskapi traditions because it served as a randomization practice that enabled hunting to be more successful over the long run. Scapulimancy appears to play a critical role in sustainable subsistence practices.

We recognize that, like the Naskapi, the Rapanui people had reasons, undoubtedly many reasons, to build and move massive statues, and we're not saying that a desire to limit childbirth and overpopulation was one of them. But we do think that the limitation of childbirth was a desirable consequence.

It's also important to make the point here that we're not saying that all of the islanders engaged in limiting the number of their offspring. Conformity across the whole population would not have been required to produce the bet-hedging benefit. Within any population there will of course be different kinds of behaviors. Some will take advantage, for example, of short-term gains, believing that it's better to do so. In terms of childrearing, this might mean that people would try to have as many children as possible. Perhaps those people are confident that there will be enough resources to support them and their children in the future. But inevitably others will take a more conservative approach. Which of these strategies will be the most advantageous over the long run, in evolutionary terms, will depend on what actually happens in terms of available resources. Each choice, wittingly or not, will affect future reproduction and long-term success. Even doing nothing affects success in reproduction because it means lost opportunities in something that might contribute to reproduction. Every action when viewed in this light has a "cost" relative to reproduction, either contributing directly to it, as in the case of food production, or taking away from reproduction when activities make no direct contribution.

At the level of a whole society, no one set of individuals has to figure out the "right" answer. Individuals and the families they form will have their own traditions, decisions, and responses to what they perceive is the appropriate choice at the time. So in terms of a practice like any form of cultural elaboration, individuals will likely vary in their level of investment. But over the long run, where conditions favor bet-hedging, whatever investments in cultural elaboration are made will be beneficial to the overall society.

Ancient Rapa Nui, a small, isolated island with limited and fluctuating food resources, is exactly the kind of place we would expect to see bet-hedging strategies take hold. And we would expect to see any activity with the effects of bet-hedging provide benefits relatively quickly since the population density would have quickly increased to a point where resources were limited.

We have seen, for example, that even growing crops in rock mulch gardens and the use of *manavai* turned out to be forms of bet-hedging. Recall that rock mulch tends to cool the soil, somewhat reducing productivity. So the islanders made a trade-off, increasing the reliability of returns while sacrificing optimal productivity at any given point in time.

Living in a dispersed community also reduced the risk in food production by growing crops from as wide an area of gardening plots as possible. Crops successfully grown in one part of the community could balance failure in another. This distribution also comes at a cost to efficiency, since aggregating farming in one place makes it more productive in the short run. Again, this was a trade-off worth making to ensure consistent and stable food returns.

An additional line of evidence for prehistoric bet-hedging on Rapa Nui comes from the curious ratio between the numbers of males and females. The earliest European explorers consistently noted that the native men of the island substantially outnumbered the women.[6] From these firsthand accounts, however, it is difficult to determine whether the observers simply didn't have an adequate sample to assess the ratio. As Cook speculated in 1774, it is possible that women were hiding or being sequestered in caves. Somewhat more direct evidence can be found in prehistoric skeletal remains. In a sample of 329 prehistoric individuals from the island, physical anthropologist Vincent Stefan identified 141 females versus 188 males, a difference of 14 percent.[7]

In general, sex ratios in populations should be roughly equal. In circumstances where there are more females than males, for example, newborn males will have more opportunities to mate than females. Thus parents who produce males will have more grandchildren than females. The propensity for male births will increase and males will become more common. A similar scenario occurs when there are more females than males. No one sex has an absolute advantage over the long run. Ultimately, the benefits balance themselves and the sex ratio generally sticks to approximately 1:1. In any actual population there will be differences given historical conditions and slightly different survival of one sex or another. In humans, for example, it is typical for sex ratios to be biased slightly toward males, with an average of 105 boys born for every 100 girls. In the case of Rapa Nui, the number of men versus women was much greater than we might otherwise expect.

The large sex ratio disparity of Rapa Nui points to an additional area of prehistoric behavior with bet-hedging consequences. Decreasing the number of women relative to men has a direct impact on reproductive rates for populations, since women are ultimately the limiting factor in the numbers of births. The reduction of birthrate would allow for greater investment in the fewer children present, providing better assurance that they would reach adulthood. Exactly how such a skewed sex ratio emerged, however, is not clear. While there is no direct evidence, it is possible that female infanticide or other mechanisms explain the sex ratio. Archaeological evidence for infanticide is difficult to evaluate, particularly since the remains of infants are rarely preserved. Nevertheless, evidence of a male sex ratio bias, however it occurred, reveals another form of bet-hedging that contributed to long-term survival.

THE BET PAID OFF

The archaeological evidence from Rapa Nui points to a delicate balance maintained between the food resources and the size of the population. A study of prehistoric skeletal remains from 125 individuals conducted by Belgian physical anthropologist Caroline Polet has documented that prehistoric islanders were reasonably healthy, though they did suffer through periods of dietary shortfalls.[8] Nutritional deficiencies can cause individuals to develop vertical and horizontal grooves on their teeth, a condition known as enamel hypoplasia and caused by disruptions in the growth of tooth enamel. On Rapa Nui, they appear in the teeth of 18 percent of the individuals studied. In addition, Polet showed that the skeletal evidence from Rapa Nui also includes incidents of cribra orbitalia, or the development of pores in the upper bone surface of the eye sockets. Cribra orbitalia is a strong indicator of anemia, a disorder brought on by decreased red blood cells and often caused by iron deficiency in the diet. This condition was found in 28 percent of the subadult individuals and 6 percent of the adults.

Together these skeletal markers indicate that the population occasionally suffered from food shortages, but also that these dietary problems were not particularly acute. Remarkably, Polet's study shows that the level of enamel hypoplasias and cribra orbitalia among prehistoric Rapanui was lower than that found on much larger and more verdant islands, such as those of Tonga and

Hawaii. The prevalence of stress indicators on Rapa Nui turned out to be lower than among the populations of medieval Europe.

On average, the prehistoric population of the island was relatively healthy, though the specter of food shortages remained close. It would seem that whatever bet-hedging strategies the islanders pursued resulted in food production that was capable, if just barely, of consistently meeting the basic needs of the population. We believe that monument construction was an important element in maintaining this balance, and that those who invested in making and moving statues did better in the long run than those who did not. Over time, as population density increased, the benefits increased, and the record shows that as this happened, greater investments in *ahu* and *moai* were made.

The record is striking. Based on admittedly poor dates, it nonetheless appears that *moai* and *ahu* construction began soon after initial colonization of the island. Early forms of statues were quite variable, and early *moai* tended to be smaller and made from a wider array of materials than later statues. Over time, the maximum *moai* size increased, and it seems the last statues erected on *ahu* were among the largest ever to have been moved. Statues such as Paro, the tallest statue ever moved to an *ahu*, weigh about 82 tons and are about 32 feet high. The *pukao* that once crowned Paro is 6 feet across and 5.5 feet high and weighs about 11.5 tons. While slightly shorter, one of the statues transported and placed on Ahu Tongariki weighed about 86 tons. And a 69-foot-tall statue left unfinished in the quarry would have weighed as much as 270 tons, had it ever been completed.

Again, we are not saying that this increasing investment in statues was a conscious strategy of bet-hedging, but it makes sense that it was of great benefit to the long-term survival and success of the prehistoric Rapanui.

But neither the individual benefits to be derived from signaling nor the fact that statue making turned out to be beneficial for survival explains why so many individuals were willing to devote so much time to it. While individuals who participated surely gained direct benefits that would have contributed to their own status and life success, there must also have been some benefit to cooperation at the level of groups of individuals. Indeed, making and moving a *moai* to an *ahu* was a community effort. Of course, many facets of our lives depend upon shared group-level behavior and, in general, humans often act for the good of groups. But from a strictly cost/benefit, rational analysis perspective, the choice to cooperate often makes little sense. If one can gain the benefits of the behavior *without* investing the labor and time to make it happen, why bother? This is known as the free rider problem. Of course, people's cooperation may have been due to the cultural beliefs about the statues and their value and an ethic about doing one's part, or even a desire to do so. We can never know about that. But we can find compelling insight about why this cooperative behavior would have emerged, in another realm of explanation, again that of evolutionary theory.

EVOLUTION OF COOPERATION

Explaining how behavior beneficial to all could emerge in societies has been the subject of much attention by evolutionary biologists, ecologists, and social scientists for at least the past fifty years. Often this discussion has focused on the conditions under which individuals might act altruistically rather than selfishly, as Darwinian evolutionary theory had seemed to predict. The earliest explanations for altruism centered largely on "kin selection," meaning that we extend altruism toward those who are related to us, and the closer the relationship the more altruistic we are willing to be. This is because by benefiting them we are also benefiting the survival of our genetic group. For example, one would expect more altruism between siblings than cousins.

But a powerful alternative to the kin-selection-based explanation for altruism is so-called group selection, which asserts that individuals benefit significantly from social, economic, or other forms of cooperation in groups. And so beneficial is this cooperation that groups with greater numbers of altruists will, under particular conditions that place a premium on that cooperation, outcompete those with fewer altruists. Critics of this idea are quick to point out that the selfish individuals within groups will always outcompete the altruists, and so over time the cooperation will break down. The exception would be groups made up of members who are closely related, which brings us back to the logic of kin selection.[9]

More recently, however, evolutionary biologist David Sloan Wilson and philosopher Elliott Sober[10] have proposed a new formulation of group selection, based on a model originally established by evolutionary biologist John Maynard Smith,[11] that addresses this issue. Their argument reasons that although it's true that altruistic individuals can be at a disadvantage to those who are selfish, if there is a reason that cooperation by individuals contributes significantly enough to success for the whole group, then group cooperation will be selected for, overriding the free rider problem. But they stipulated two conditions that are necessary in order for this to occur. First, the overall population must be divided into cohesive subgroups in which the members predominately interact and compete among the members of their own group, rather than with those of the other groups, which would lead to intergroup conflict. And second, individuals from these groups must not be entirely isolated from one another; they must mix and interact to some degree. This interaction might occur with periodic gatherings of all the groups into a single population for some occasions, such as ceremonies. Or it might occur through intermarriage or the exchange of individuals between groups.

The archaeological record of Rapa Nui, as described in the last chapter, suggests a society that was structured in just this way. The various communities operated for the most part as independently functioning groups, each with certain variations in stylistic details of statue forms, *ahu* shape, *umu* construction, and other artifacts. The communities were also somewhat biologically isolated, but there are signs of some degree of interbreeding. So some of the latest work in evolutionary theory offers a good explanation of why cooperation within these groups would have prevailed; doing so was to the advantage of the group as a whole and so was selected for over time.

This brings us to the issue of the culture's collapse. How did the delicate balance of environmental stewardship and social cooperation that we've described get out of whack? The collapse of the culture was surely dramatic; about that there can be no dispute. But as we'll explore in the next chapter, it was not a case of the islanders' committing ecocide; it was a consequence of European arrival on their shores.

CHAPTER 9

The Collapse

It is a sad fact that in these islands, as in North America, wherever the white man establishes himself the aborigines perish.

—Commander H. V. Barclay,
HMS *Topaze,* 1868

The island was baptized with the blood of its children; and, as though this massacre had been an omen for the future, it was the scene, in the middle of last century, of one of the most hideous atrocities committed by white men in the South Seas.

—Alfred Métraux, 1957

Captain Roelof Rosendaal and the crew of the Dutch vessel *De Afrikaansche Galey* (The African Galley) spotted the island of Rapa Nui on the afternoon of Easter Sunday, April 5, 1722. Rosendaal waited until the other ships in Captain Roggeveen's fleet (the *Arend* and *Thienhoven*) caught up with the *De Afrikaansche Galey* before the three vessels began to approach the island. Seeing smoke rising from the island and people onshore, the three captains decided that in the following morning two of the better-armed ships, the *Arend* and *Thienhoven,* would undertake an expedition ashore. Their intent was to "show all friendliness towards the inhabitants, endeavouring to see and inquire what they wear or make use of either as ornaments or for other purposes, also whether any refreshments in the way of green stuff, fruit, or beasts can be procured by barter."[1]

As they prepared to make landfall, a native Rapanui man paddled a small canoe out to the *Thienhoven*. This man examined the ship closely, testing the tautness of the spars, the rigging and guns, as well as his own face in a mirror. The Dutch did the same, noting that he was "quite nude" and "hapless." A short while later, the man left the ship after accepting two blue glass beaded necklaces, a small mirror, a pair of scissors, and an assortment of other items the Dutch had available for gifts and trade. Not long after this initial visit, a "great many" canoes approached the ships and their occupants swarmed aboard. These visitors were curious about the ships, but were particularly interested in obtaining material goods from the Dutch. Roggeveen wrote:

> These people showed us at that time their great cupidity for everything they saw; and were so daring that they took the seaman's hats and caps from their heads, and sprang overboard with the spoil. . . . There was also an Easter Islander who climbed in through the cabin window of *De Afrikaansche Galey* from his canoe, and seeing on the table, a cloth with which it was covered, and deeming it to be a good prize, he made his escape with it there and then; so that one must take special heed to keep close watch over everything.[2]

This would not be an isolated incident. On their first landing on the island, the Dutch were surrounded by local people who greatly desired European goods. Rapanui grasped the muzzle of the European guns and attempted to strip the soldiers of jackets. As the struggle grew more frenzied, the Dutch resisted and the situation began to threaten the isolated group of sailors. Rapanui "picked up stones, using threatening gestures as if to pelt us with them."[3] Out of fear and misunderstanding, the Dutch soldiers opened fire on the crowd. The Dutch estimated that ten or twelve Rapanui were killed and some unknown number were wounded. Fortunately, the Dutch commanders called for a cease-fire and the crowds of Rapanui fled in terror.

Such scenes were replayed over the decades to come with steadily increasing amounts of violence. While we are probably inclined to see these conflicts as caused by imperialist Europeans, their cause was rarely one-sided. In particular, the Rapanui demonstrated a passionate interest in exotic goods that arrived with the foreign visitors. For the first century of interaction between islanders and outsiders, almost every early explorer, missionary, trader, and tourist remarked how the islanders were prepared to borrow, beg, and simply take anything they could get their hands on.

Curiously, one item that apparently had great value among the islanders was hats. Over and over, accounts of visitors remark how islanders would relieve them of hats when the chance arose. In 1774, for example, Cook remarked that "some of the gentlemen also made an excursion again in the evening, with the loss only of a hat, which one of the natives snatched off the head of one of the party."[4] Cook even noted the presence of hats and other European items that were acquired from the unlucky members of previous expeditions: "Before I sailed from England, I was informed that a Spanish ship visited this isle in 1769. Some signs of it were seen among the people now about us; one man had a pretty good broad brimmed European hat on; another had a grego jacket; and another a red silk handkerchief."[5]

Remarks about the islanders' penchant for hats and other European bits of clothing continue in the accounts of nearly every subsequent visitor. Several years later, for example, during his visit in 1786, La Pérouse also noted this passion for hats. While he had brought a variety of material to present to the island inhabitants as gifts, some items were prized over all others. In this passage La Pérouse is frustrated by the lengths to which the islanders will go to acquire a hat, even staging events to distract the French sailors with bawdy dances.

> I made them several presents, and found that they preferred small remnants of printed cotton, about half a yard long, to nails,

> knives, and beads but they were still fonder of hats. Not only did islanders snatch hats when the opportunity presented itself but specific schemes were enacted to lure the crew of the *Astrolabe* and *Boussole* into position for hat acquisition. Groups of women salaciously presented themselves to groups of sailors. While entranced, the sailors were relieved of their headgear. The Indians solicited us to accept their offers, and some among them gave us an exhibition of the pleasures they were capable of affording. The agents in these transactions were not otherwise concealed than by a simple covering of cloth, the manufacture of the country. While our attention was attracted to these tricks of the women, our hats were taken from our heads, and our handkerchiefs stolen out of our pockets.[6]

This scene struck La Pérouse to such a degree that it was one of the few sketches included in his written account of his daylong visit.

Figure 9.1. Islanders distracting members of La Pérouse's expedition in order to acquire European hats and other personal items.

In fact, even though La Pérouse's visit lasted less than a day, it was not long before nearly every member of his expedition had lost his hat, leaving them frustrated and their heads subject to the heat of the sun. A second passage and drawing describe La Pérouse's lament: "An Indian who assisted me to descend from a platform, after rendering me this service, took away my hat and fled with the utmost speed, followed as usual by all the others. I did not suffer him to be pursued, not being desirous of the exclusive privilege of defense from the sun, as we were almost all without hats."[7]

The passion displayed by the Rapanui for European goods was more than just the effect of curiosity and the lack of a sense of personal property, as early European visitors postulated. In fact, we can trace a deep tradition of headgear as symbols of status in the prehistory of the island. Early drawings of islanders made during Cook's voyage show individuals wearing elaborate hats made of feathers. And we see hats prominently displayed as decorative adornments for the *moai*. These hats—*pukao*—were themselves massive monuments that involved impressive feats of engineering and human labor.

Large, squat cylinders, *pukao* are carved out of red scoria, which is coarse, porous, and cindery lava. Its distinctive red color comes from the oxidation of iron within the stone. While there are a variety of places on the island one can find quantities of red scoria, it appears that the vast majority of *pukao* were quarried out of a small crater located on the western side of the island and north of Rano Kau. This crater, known as Puna Pau, is a small vent of the Terevaka volcano that erupted around 220,000 years ago.

We have identified and mapped nearly one hundred *pukao* on the island. They range in size from just a few feet in diameter and height to as large as eight feet in diameter and over eight feet high. And they weigh many tons. The *pukao* appear to have been moved by rolling, though we have only limited direct evidence for this. Their paths took them from a quarry where they were mined to destinations at *ahu* located across the island.

The location of the *pukao* quarry at Puna Pau has been known since the earliest visitors to the island.[8] In March 1774, Cook reported the description of the quarry by a small group that made an excursion into the interior parts of the island.

> In a small hollow, on the highest part of the island, they met with several such cylinders as are placed on the heads of the statues. Some of these appeared larger than any they had seen before; but it was now too late to stop and measure any of them. Mr. Wales, from whom I had this information, is of the opinion that there had been a quarry here, whence these stones had formerly been dug; and that it would have been no difficult matter to roll them down the hill after they were formed. I think this is a very reasonable conjecture and have no doubt that it has been so.[9]

Once the big wheels of red scoria were rolled to *ahu* they were then shaped into a variety of *pukao* hat forms. Though in general they were roughly cylindrical, some were more conical, and the record indicates that they were shifted slightly forward when they sat on the heads of *moai.* The chief pilot of the Spanish frigate *Santa Rosalia,* who sailed with González in 1770, wrote an intriguing description of *pukao.*

> The diameter of the crown is much greater than that of the head on which it rests, and its lower edge projects greatly beyond the forehead of the figure; a position which excites wonder that it does not fall. I was able to clear up this difficulty on making an examination of another smaller statue from whose head there projected a kind of tenon, constructed to fit into a sort of slot or mortise corresponding to it in the crown; so that by this device the latter is sustained notwithstanding its overlapping the forehead.[10]

It is in the islanders' intensity of interest in the hats of the Europeans who arrived on their shores that we see just one aspect of a dramatic shift in the culture of the island, away from their focus on the statues as signals of prestige and toward the goods of all kinds brought by the Europeans. Some five centuries of tradition of *moai* construction ended nearly overnight.

Historic accounts indicate that the collapse of prehistoric *moai* traditions happened with remarkable speed. Some statues were already down by Cook's visit in 1774, just a few short years after the Spanish arrival in 1770. By 1804, Russian sailor Yuri Lisjanskij reports seeing just twenty standing *moai.* In 1830, British sailors aboard the HMS *Seringapatam* note just eight remaining upright. The last report of standing *moai* comes in an account in 1838 by French naval officer Abel Aubert Dupetit-Thouars. At that time there were just four statues noted on the west coast of the island. By 1868, J. Linton Palmer, a British naval surgeon aboard the HMS *Topaze,* remarked that no statues remained standing on the island's *ahu* platforms.[11]

In some cases toppling of statues may have been purposeful, but many more likely came down as a result of inattention and lack of maintenance. What is clear, however, is that with the arrival of Europeans the rationale for participating in *moai* construction and movement had been undermined; the activity had lost its value. The acquisition of European goods became the new form of obtaining prestige and the islanders demonstrated that they would go to almost any length to acquire these items. For example, the historical record shows that the islands' men began to offer women to visiting European sailors in exchange for trinkets and other items. In addition there are accounts of elaborate schemes for tricking the Europeans, such as rocks being stuffed into bundles of sweet potatoes to make the bundles look as though they were bountiful. Dances were held to distract Europeans so that objects could be stolen. Immense attention was focused on acquiring goods. European hats, among other foreign novelties, had become the new costly signals communicating access to new wealth and technology.

It is during the early period of historic contact that we find evidence of Rapanui engaging in what anthropologists have called "cargo cults." Cargo cults are cultural traditions that occasionally appear after contact between populations who use widely divergent technology. In these situations, ritual behavior is invoked by one group in hopes that the other will return and bring with them social change or foreign goods. This behavior can take the form of reenacting foreign behavior as an attempt to "attract" visitors (and their goods) back to the island. For example, cults emerged in islands across the southwest Pacific after Japanese and American troops spent time on previously isolated islands during World War II. On islands such as those of Vanuatu, in efforts to bring back American servicemen, locals constructed airstrips, wooden control towers, fake radios, and even bamboo airplanes. Often these life-size models were "manned" to simulate the conditions that the islanders had seen when Americans had occupied the island during wartime.

On Rapa Nui a similar kind of activity followed European contact. In a number of locations, islanders constructed what are referred to as *miro-o-one,* or "earth ships," as well as *hare-a-te-atua,* a ship-shaped house. These are monuments used as locations for islanders to enact a kind of show in which individuals imitated the activities, postures, and language of European sailors. Katherine Routledge described the shows that took place in these structures based on accounts provided by her local informants.

> The simplest form of this celebration took place on long mounds of earth known as the "miro-o-one," or earth-ships, of which there are several in [*sic*] the island, one of them with a small mound near it to represent a boat. Here the natives used to gather and act the part of a European crew, one taking the lead and giving orders to the other. A more formal ceremony was held in a large house. This had three doors on each side which the singers entered, who were up to a hundred in numbers and ranged themselves in lines within.[12]

One aspect of this tradition consistent with other cargo cults was the belief that the structures represented the ships that brought "the gods." When Routledge inquired about these gods in 1914, she was told that they were "the men who came from far away in ships. They saw they had pink cheeks, and they said they were gods."[13]

In 2002, when we surveyed interior areas of the island as part of our early fieldwork, we discovered the site Routledge had described. This *miro-o-one* is still visible as a mound formed by earth roughly the size and shape of an eighteenth-century European ship. A depression around the mound probably once formed a ditch that, when filled with water, would have provided the appearance of the ship floating at sea. The similarity in the dimensions is not surprising, as historic accounts describe islanders boarding ships in order to make careful measurements of visiting vessels. La Pérouse, for example, writes: "The care they took to measure our vessel convinced me, that they had not contemplated our arts with stupidity. They examined our cables, our anchors, our compass, and our steering wheel; and in the evening they returned with a string to take their measure over again; which showed that they had had some discussion upon the subject on shore, and that doubts had remained in their mind."[14]

The abandonment of long-standing *moai* traditions was only one of many elements that changed with the introduction of Europeans into the social systems that had evolved on the island over hundreds of years of isolation. The Europeans, however, were largely unaware of the impact they had on the delicate balance of social and economic life that had been achieved on the island.

Some of that ignorance was due to the fact that most early visitors remained on the island for just a short time. The Dutch, for example, were ashore in 1722 for only about a day. On April 11, they encountered rough seas and an anchor rope snapped; then, as the fleet pushed dangerously close to the island's cliffs and rocky shore, a second rope broke. Captain Bouman of the *Thienhoven* wrote of being nearly shipwrecked. The next day, on Sunday, seven days after first sighting the island and with only one day ashore, the Dutch departed under perilous conditions. By six in the evening Easter Island was more than six miles in the distance as Roggeveen and his small fleet sailed west toward the Tuamotu Archipelago, largely oblivious of the impact their short visit had had.

Thus Rapa Nui's collision with the outside world began with ignorance and mutual misunderstandings. For the Rapanui, everything changed in those few fateful days. A sad and ironic footnote to the story: Dutch Sergeant Major Carl Friedrich Behrens noted "there were many shot dead here, among them also the man who had earlier been by us, which grieved us sorely."[15] Of course, he refers to the naked man who first ventured far offshore in a small canoe and boarded the ship to welcome these strange foreigners to their island world.[16] The story, however, would get worse, much worse.

Within three days of their visit ashore, the Dutch had sailed over the horizon ignorant of so much of what had transpired. Indeed, they were ignorant of even the possibilities. Germ theory, discovered in isolated fits and starts, was not widely understood until the mid- to late nineteenth century. So, unintentionally and unaware, the first European visitors introduced diseases that initiated hideous and devastating germ warfare upon the island. It happened in at least two ways. First, it would be naive to think that venereal disease was not part of the Dutch exchange—and certainly it arrived with subsequent visitors. Second, other pathogens of little consequence to the Europeans who harbored them had reached the island with the sailors. Either way, hordes of new germs were unleashed, posing a far greater and devastating threat than the muskets that had killed a dozen or so of the Rapanui assembled on the shore just days earlier. What happened next was witnessed only by the victims, the Rapanui themselves.

Jared Diamond aptly calls Old World diseases the "lethal gift of livestock," since the big killers such as smallpox, influenza, tuberculosis, plague, measles, and cholera evolved in herd animals and then jumped species to densely populated farmers in the Old World.[17] For comparison, Ebola, dengue fever, and HIV are more recent examples of animal-to-human transfers.[18]

Evolutionary biologists have shown that populations who have survived repeated waves of disease tend to form a kind of détente with the most deadly varieties.[19] Mutual benefits for hosts (a person) and the pathogen bring this about. Pathogens need hosts to grow, but doing so too aggressively can cause death to the host. Once an infected person dies of a disease, the responsible pathogen will usually also perish and thus cannot infect others. So the longer an infected host lives, the more chances the pathogen has to spread to new individuals. And the more lethal the pathogen, the more rapidly it must be transmitted from one host to another.

When it becomes more difficult for a pathogen to jump to a new host due to population loss or the establishment of barriers for infection (such as isolation of sick people), evolution tends to cause the virulence, or deadliness, of a pathogen to decrease since only those pathogens that are able to get passed from individual to individual survive. Those variants that kill quickly cannot be maintained in the population. Thus over time diseases become common, yet relatively mild. At the same time human populations evolve in reaction to pathogens. The individuals who survive the effects of pathogens are those with immunity or a tolerance to the effects of the disease.

Consequently, given their history and continuous exposure to diseases that jumped from domesticated animals, Old World populations tend to carry consistent but largely nonlethal levels of disease with them. These diseases were persistent, but presented only relatively minor inconveniences.

But Native Americans, Australians, and the populations of the Pacific, in specific, had few if any domesticated—especially herd—animals, and thus had limited immunity to Old World pathogens. When those pathogens invaded, the consequences were unimaginable. Described as the "American Holocaust," contact from the Old World to Native American populations led to epidemics where more than 90 percent of the population died. In the New World and the Pacific, native peoples sustained huge losses following initial contacts, but before literate witnesses recorded the horror.[20] It would be difficult to explain how the impact of disease would not happen with the Dutch and subsequent arrivals on Rapa Nui. In each case hasty departures made the impact of epidemic or lingering diseases effectively invisible to Europeans, who were largely unaware of disease causes and consequences for the island.[21]

Such was the case on Rapa Nui. Epidemics would claim a great proportion of the island's population. Roggeveen's impression of a few thousand people on the island in 1722 may well have been accurate, but the population likely soon collapsed to just a few hundred survivors, perhaps in only a few years. The precise numbers are anyone's guess.

A catastrophic population collapse would have obliterated the social, political, and economic status quo. The devastating epidemics and longer-term effects of venereal disease, for example, would have come as a shock to the Rapanui.

If by about 1725 there were only a few hundred survivors left on Rapa Nui, they effectively formed a new founding population—of survivors. The Rapanui people were isolated again from the outside world and its diseases for forty-eight years thereafter, and this small group would have rebounded relatively quickly, approaching the original population size, probably within three to four generations.

But in 1770 the onslaught began anew. As part of Spain's effort to discover and colonize new lands, especially the rumored Southern Continent, an expedition of two warships led by Felipe González de Haedo set sail from the port of Callao in Peru. The Spanish were interested in monitoring colonial efforts of the British and French on the islands and coasts of the South Pacific. In particular their goal was to locate Davis Island, supposedly sighted by English buccaneer Edward Davis in 1687, or Easter Island, if indeed it was one and the same, and claim it before the British could do so. After five weeks at sea, González and his crew arrived on Easter Island on December 15, 1770. True to their mission, the Spanish claimed the island in the name of King Carlos III and renamed it Isla de San Carlos. González made it official; he asked the island chiefs to put their mark on the Spanish document ceding their land to foreign rule. Three of the chiefs drew symbols on the document. González and his men would spend six days ashore on the island.

The notes from the expedition indicate that as the ships approached the island, it appeared "quite covered in greenery as far down as the sea-beach, showing the fertility of the country."[22]

Before anchoring the next day, the Spanish saw more than eight hundred people gathered along the shore and nearby heights. In the log: "there was not the least appearance of hostility, nor of the implements of war about them; I saw many demonstrations of rejoicing and much yelling."[23] Repeating a scene from forty-eight years earlier, islanders swam and paddled small canoes out to meet the Spanish ships. They clambered about the ship and danced when the sailors played bagpipes.

Of the Rapanui, the Spanish note that "it appears as if among themselves their goods are held in common," going on to record:

> The women go to length of offering with inviting demonstrations all the homage that an impassioned man can desire. Nor do they appear to transgress, in this, in the opinion of their men; for the latter even tender them by way of paying us attention. . . . It can only be inferred that the women whom we saw are held in common among them, although we noticed that the older and more important men retain some preference in the matter. . . . The women we saw were much fewer in number than the men.[24]

The Spanish recount in detail a ceremonial planting of three crosses on the three cinder cones that form an alignment on the northern flanks of Poike. The Spanish flag was hoisted and the troops brought to attention—all part of the rituals of signing over possession of the island—and everyone cheered the king of Spain seven times before a triple volley of musket fire followed by a twenty-one-gun salute from the ships. The following day the Spanish departed, feeling they had accomplished their mission.

While no violence erupted during the Spanish visit, something more potent had been unleashed on Rapa Nui. From the six days on the island and the explicit references to sexual encounters, we can be certain that venereal disease gained a foothold. Epidemic disease must also have followed the Spanish visit, and this time unwitting witnesses of the impact would arrive on the scene within four years: the British.

* * *

Navigating the coast of the island for a safe anchorage, Cook arrived about one mile from shore at the settlement known as Hanga Roa (where the island's only town is located today). Two men approached in a canoe offering a cluster of bananas, which was quickly hoisted by rope onto the ship. The crew was badly malnourished, and Forster recorded "the sudden emotions of joy in every countenance at the sight of this fruit are scarcely to be described."[25] About this exchange, Cook wrote, "this gave us a good opinion of the islanders, and inspired us with hopes of getting some refreshments, which we were in great want of."[26] Not finding a better anchorage farther along the west coast, Cook returned to Hanga Roa and dropped anchor. Without hesitation, an islander swam to the ship and came aboard. "Maroowahai," as his name was recorded, stayed on the ship two nights and a day conversing as much as possible with a Tahitian named Mahine, who served as translator on the voyage. Cook notes that the Rapanui language was "wholly unintelligible to all of us."[27] Viewing the island from sea, Forster observed: "On the slope we discovered several plantations . . . but the surface of the isle in general appeared to be extremely dreary and parched, and these plantations were so thinly scattered upon it that they did not flatter our hopes of meeting with refreshments. . . . We could easily perceive that there was not a tree upon the whole island, which exceeded the height of ten feet."[28]

The next morning, March 14, Cook went ashore "to see what the island was likely to afford us."[29] Hundreds assembled on the beach to see the visitors. Cook, still suffering the effects of an unspecified illness, stayed behind near the shore, where he was accompanied by Maroowahai, Mahine, and many islanders. From his interactions onshore, Cook soon concluded, as he wrote in his journal, that "They were as expert thieves, and as tricking in their exchanges, as any people we had yet met with. It was with some difficulty we could keep the hats on our heads; but hardly possible to keep any thing in our pockets, not even what themselves had sold us; for they would watch every opportunity to snatch it from us, so that we sometimes bought the same thing two or three times over, and after all did not get it."[30]

Observing their first day ashore, Georg Forster reported:

> We saw but few arms among them; some however had lances or spears, made of thin ill-shapen sticks, and pointed with a sharp triangular piece of black glassy lava. . . . One of them had a fighting club, made of a thick piece of wood about three feet long, carved at one extremity; and a few others had short wooden clubs, exactly resembling some of the New Zealand *patoo-patoos*, which are made of bone. We observed some who had European hats and caps, chequered cotton handkerchiefs, and ragged jackets of blue woollen-cloth, which were so many indubitable testimonies of the visit which the Spanish had made to this island in 1770. . . . The number of women in the crowd did not exceed ten or twelve.[31]

That night, after returning to the *Resolution*, Georg Forster described natives who had swum to the ship, anchored some three-quarters of a mile offshore. He wrote in his journal: "They expressed the most unbounded admiration at everything they saw, and every one of them measured the whole length of the vessel from head to stern, with his extended arms; such a great quantity of timber of so stupendous a size being altogether incomprehensible to people whose canoes were patched of many small bits of wood."[32]

Among them was one woman who carried on a particular traffic of her own. The accounts suggest that sexual encounters were common between local women and the British sailors. The following day, Cook dispatched Lieutenants Pickersgill and Edgecumbe, along with Johann Forster, William Hodges, and the ship's doctor, to reconnoiter the interior of the island. Cook again stayed behind and engaged in trading with the Rapanui. On their trek across the southern interior of the island, they observed the volcanic landscape, gardens, and of course, the giant statues. About a hundred Rapanui, including only four or five women, joined the march. At one point they stopped for Hodges to draw some of the monuments, where they also saw an entire human skeleton on the ground's surface—most likely direct evidence of the effects of recent epidemic disease that came with the Dutch or Spanish.

Unfortunately, it was at this point that an islander made off with a bag containing some nails that had been left by one of the sailors. Lieutenant Edgecumbe witnessed the act and fired his musket loaded with small shot. The wounded Rapanui man threw down the bag but he fell soon after. Forster wrote of the incident, "though this was the only instance of firing at a native during our stay at Easter Island, yet is to be lamented that Europeans too often assume the power of inflicting punishments on people who are utterly unacquainted with their laws."[33]

As they continued on their tour, the crew was directed to a deep well that contained fresh, potable water, where they "drank heartily" and potentially introduced disease to at least one source of drinking water. Evidence of recent abandonment of statue construction was also noted. Farther along, the group passed "several large statues, which had been overturned." On their way back they returned to a large statue the natives called "Mangototo," "and in the shade of which we dined." The crew also recorded that "in its neighborhood we met with another huge statue, which lay overturned; it was 27 feet long, and 9 feet in diameter, exceeding in magnitude every other pillar which we had seen on the island."[34]

On March 17, 1774, the British departed after three days ashore. In his journal, Georg Forster quotes Mahine's succinct parting comment in Tahitian about Rapa Nui: "*tata maitai, whennua eene* (*ta'ata maita'i, fenua 'ino*), that the people were good, but the island very bad. . . ."[35] Cook's view of this humble island was clearly shaped by its inability to serve his desperate needs as well as his lack of understanding of its recent history.

At the time of Cook's visit, just four years after the Spanish had arrived, the Rapanui were almost certainly suffering in the aftermath of disease outbreaks. Indeed, the 600 to 700 people Cook reports for the island's population must be survivors of whatever pathogens were introduced and epidemics that ensued. And there are venereal diseases; they don't kill quickly, but linger to inflict long-term maladies, reproductive problems, and sterility. Numerous accounts describe instances in which women were offered to sailors, in exchange for foreign goods or even as distractions. Forster, for example, observed that the women "were neither reserved nor chaste, and for the trifling consideration of a small piece of cloth, some of our sailors obtained the gratification of their desires."[36] La Pérouse indignantly complained about his crew being offered young girls, thirteen or fourteen years old. He was, however, quick to clarify: "None of our people availed themselves of the barbarous right thus attempted to be conveyed to them; and if certain moments were devoted to nature, the desire and consent were mutual, and the women made the first offers."[37] In short order venereal diseases became common on the island. By 1830, ship captains expressed concern about landing on the island due to the prevalence of syphilis.[38]

The British were unknowing witnesses to an epidemic that had just passed. But what had just transpired was still incomprehensible to them. They were perplexed by the small population size, what they perceived as poverty, and generally the disheveled state of things; in hindsight, this is precisely what the aftermath of epidemic and a population crash would look like.

Once the location became well-known, Rapa Nui became a common landing site for whalers, sealers, and other hunters seeking provisions, including fresh water and produce such as sweet potatoes, yams, and bananas. But from 1776 to 1864—for seventy-eight years—when the vast majority of visitors were whalers, little, if any, information is recorded of their stays. There were at least fifty ships visiting or sighting the island in the first half of the nineteenth century, but there were probably many more that escaped any documentation. The motives for visiting grew increasingly malicious.

In 1805, the *Nancy,* captained by the ruthless J. Crocker of Boston, launched a bloody battle against the Rapanui and enslaved twelve men and ten women as "laborers" for a seal-hunting operation in Más Afuera (Alejandro Selkirk Island in Chile). The islanders were kept in shackles belowdecks. Then, after sailing three days from Rapa Nui, over two hundred miles away, the captives were allowed on deck, where the men promptly jumped overboard into the open ocean; the sailors stopped the women from doing the same. Captain Crocker sent a whaling boat back in an unsuccessful attempt to recapture them, but the Rapanui men evaded the sailors by diving under the water. The *Nancy* continued with the women only, but reportedly returned to Rapa Nui several times to kidnap more islanders for slavery.

Other ships attempted to land at Rapa Nui, but understandably were met with hostility great enough to drive them off. By 1816, Rapanui attitudes were still justifiably hostile when a Russian expedition led by O. E. Kotzebue arrived on the island. When Kotzebue and his crew of seventeen men tried to go ashore, they were pelted with stones thrown by the Rapanui, who laughed and shouted at them. In response to the stone throwing, the Russians fired shots and forced their way onshore. As the Russians began to engage in trading bits of iron and knives for vegetables, they were faced with more stone throwing and hastily departed. It seems the Rapanui had decided "visitors" were no longer welcome.

By the early nineteenth century the whaling trade was in full swing in the Pacific. As the great whales of the North Atlantic became increasingly scarce, hunted to near eradication, whalers began to ply the Pacific in search of whales, despite the long distances and multiyear voyages. Between 1820 and 1860, the primary commerce of the Pacific became whaling. During this period, Rapa Nui received many whaling ships in need of fresh provisions, crew, and "entertainment." It was common for islanders to be seized and forced to serve as crew on these vessels. Ashore and on board ships, women and young girls earned trade items for their favors. Linguist and historian Steven Roger Fischer reports that Captain Waldegrave of the HMS *Seringapatam* logged that "the women admitted the embraces of the sailors in the most unreserved manner," and Midshipman John Orlebar of the same crew reported: "We found that chastity was not in their catalogue of virtues, but certainly, proved with us, I am ashamed to say, their best article of traffic."[39] Undoubtedly disease continued to spread and ravage the population. As Fischer has aptly described, "such associations necessarily altered island life further. Although the old culture was being lost, no new replacement was being consciously fashioned. Islanders simply reacted to the series of intrusive events, creating in the process a 'jury-rigged society' beating towards an unknown horizon."[40]

It was during this time that European misconceptions about the island were formed. In 1845, a French tabloid-style journal, *L'univers,* published a report that a young commander of a French vessel landed on Easter and narrowly escaped being eaten by cannibals: "Mr Olliver was brought back on board; his whole body was covered with wounds. He had, on various parts of his body, the teeth marks of these cruel islanders, who had begun to eat him alive."[41] Most recognize this as a sensationalized hoax,[42] but it is the first known, yet dramatic, claim of cannibalism on Rapa Nui. Despite its certain fictional origins, it is probably no coincidence that the next mention comes with French missionaries, and then grows as a persistent theme in later accounts. In his review of the island's history, Fischer concludes "one thing was certain . . . Easter Islanders were anything but the hostile 'cannibals' of Pacific lore. At worst they could be incited to stone-throwing."[43] But the cannibal label would later serve colonial stereotypes and reinforce fears.

Foreign ships continued to call at Rapa Nui throughout the nineteenth century. The abuses continued, some recorded for history, others only rumored. Commercialization, as seen in Hawaii and New Zealand, didn't come to Rapa Nui, not yet. The island offered only sweet potatoes, bananas, idols, brackish water, and sex.[44] And its location, closer to South America than the other islands of Polynesia, meant the island was a target for slave raids.

By the mid-nineteenth century, the cumulative impact of multiple encounters had taken an enormous toll on Rapa Nui. But something even worse would befall the Rapanui between 1862 and 1888: large-scale slave raiding ("blackbirding") arrived from Peru. At about the time Americans had gone to war with each other over slavery, it had already been abolished in South America. In the absence of slavery or sufficient migration, Peruvian businessmen demanded cheap labor. In place of slavery per se came indentures—contracts that were sold to the highest bidders in place of the people themselves. It was a loophole that allowed traders to traffic people and continue slavery under a different name. Thus Peru began the process of granting licenses to bring "immigrant" laborers from islands across the Pacific.

In 1862 the slave raids began in earnest. Islanders were recruited with fanciful tales and promises, and sometimes just by force. Those who attempted to flee capture were either caught or simply shot. Resistance grew and Rapanui took to hiding in family caves, some concealed for protection. At the same time, the island became a staging area—the last stop before returning to Peru—for captive Polynesians, especially from the Tuamotu Archipelago, where in some cases the population of an entire atoll was taken by false assurances, by gunpoint, or perhaps eventually by both. Fischer reports that by 1863 as many as 1,500 Rapanui had been abducted or killed, as evidenced by 1,408 registered in Peru (of whom 1,054 had "signed" legal contracts of indenture).[45] These islanders were forced to serve as agricultural laborers and domestic servants, and many also ended up in the hideous guano mines of the Chincha Islands of South America. Many succumbed to one of an array of diseases: smallpox, dysentery, tuberculosis, for which they had little if any natural immunity. Hundreds died under wretched conditions.

As news spread of Polynesians from the Marquesas, Tuamotus, and the Australs—many of whom were Christian converts—as well as Easter Islanders blackbirded to Peru, the Catholic mission in Tahiti, led by Bishop Jaussen from France, petitioned for their repatriation. Jaussen, with support from the Vatican, pressured the French minister in Lima to halt this slave trading posing as "immigration." The Peruvian government was forced, under weight of international political pressure from France, to cancel the "immigration" licenses in April 1863.[46] Orders came down to assemble all Polynesian "immigrants" for their return home. But repatriation brought yet another tragedy.

At Callao, where Polynesians had been assembled, a quarantined crew from the American whaling ship *Ellen Snow* was allowed ashore. The American crew members had smallpox, and while the Peruvians received vaccinations, the Polynesians did not. Soon many were sick and dying. Those who survived and returned home unwittingly brought the disease with them, no doubt among others. A new wave of disease hit the Marquesas, Tuamotus, Australs, and Rapa Nui, and in many cases new epidemics would kill half the island's remaining population within weeks.[47]

For Rapa Nui, Fischer reports that of the approximately 1,500 Rapanui who were blackbirded to Peru, the vast majority died there. In the repatriation from South America to Polynesia, eighty-five of the survivors died at sea, leaving a mere dozen or so Rapanui who actually made it back home. Then in 1871, a majority of islanders left for Tahiti and Mangareva; and even in their neighboring islands of Polynesia, the Rapanui met with death in large numbers. By 1877, the native population on the island had reached its all-time recorded low of just 110.[48]

Through a series of disastrous encounters with foreign visitors, the Rapanui population had collapsed, rebounded, collapsed again, and then recovered to a degree, only to be ravished in slave raids. The loss of population was only one of many consequences for this collapse. Much of the cultural lore, traditional knowledge, and social practices were lost in the narrow bottlenecks that followed epidemics, and as Fischer describes, depopulation brought heated disputes over lands once held by the disappeared and deceased. "The years 1863 and 1864 were marked by intermittent hostility, persisting dearth, and continued mortality. When crops were to be allocated, violent quarrels broke out. These then escalated into larger 'tribal' conflicts that eventually resulted in 'looting, devastation and famine.'"[49]

The desperation recorded in 1863–64, we surmise, certainly would have materialized before, indeed, coming after each epidemic and the population crashes that no early European visitors stayed long enough to witness. These social conflicts, evidenced in the details of the late nineteenth century, came out of the shock and disarray of disease-induced collapse and enslavement. These were the problems invoked by a fraction of a decimated society, the survivors, vying for what they could in the cruel world they had come to know. The problems were social, not a result of environmental ruin. History is the witness that Rapa Nui suffered near genocide, not self-inflicted "ecocide."

In December 1863, a novice French Catholic missionary, Eugène Eyraud, sailed from Tahiti for Rapa Nui. Joining him were a few Rapanui survivors of the slave raids. In January 1864 they reached Rapa Nui. But Eyraud's reception upon arrival on the island was not friendly, and without warning, the ship that had delivered him promptly sailed away. Eyraud spent nine difficult months on Rapa Nui. His account reads much like a British television sitcom: a series of misunderstandings and humiliations endured by a hapless foreigner living among a wily local population skilled in depriving visitors of their belongings. Eyraud guilelessly agrees to "loan" his chickens to inquisitive islanders for safekeeping: they are quickly eaten, of course. By the end of his stay, he is destitute and wearing nothing but a blanket and an old pair of shoes. With the arrival of a French ship, Eyraud happily jumps aboard and quickly departs for Valparaiso on the Chilean mainland.

Later in 1864, with support from Valparaiso, Father Hippolyte Roussel, another Frenchman who had served in the Tuamotus and Mangareva, would lead a new mission. Eyraud ultimately returned with him along with three Mangarevans who had joined the missionary efforts on their own island.

Roussel, Eyraud, and the Christianized Mangarevans settled in Hanga Roa and rapidly began converting the islanders. The Mangarevan missionaries developed a kind of pidgin Tahitian-Rapanui (which had a great influence on the development of the modern Rapanui language), and Roussel spoke Paumotu (the Polynesian language of the Tuamotus). This made these newcomers much more welcome. The Mangarevans and Rapanui also found a common Polynesian identity complete with similar, or even assimilated, oral traditions from Mangareva, or perhaps originally shared between them.[50]

It was also with Christianization that typical caricatures and obligatory vilification of the so-called heathen or cannibal past entered the Rapanui story. In 1869, Roussel wrote that "cannibalism was practiced for a very long time and only disappeared entirely with the introduction of Catholicism. Under the sweet and charming exterior . . . the natives hid their deceitful, violent and sometimes ferocious character. *And who knows how many foreigners have been eaten?*"[51] The notion that the past was "savage," and the future "civilized," was critical to conversions and teaching the gospel. Indeed, references to cannibalism became a common theme only after the missionary presence, and these references become increasingly embellished, especially in the European accounts painting a treacherous picture of these "savages" and their wild past.

With donations to the mission from the mainland, cows, a bull, and horses arrived on Rapa Nui. Many Rapanui abandoned their households in other parts of the island and took up residence in the "town" of Hanga Roa. But an epidemic of tuberculosis was consuming the island. Rapanui old and young were dying in droves. Even native suicides became commonplace. It was around the same time, in 1868, that the Frenchman Jean-Baptiste Dutrou-Bornier began purchasing land from the Rapanui. He had big plans to acquire more land in the aftermath of epidemic and then indenture all of the remaining Rapanui to work plantations in Tahiti or elsewhere. He would get Rapanui land and remove the burdensome presence of the native population. This would leave the island empty to exploit as a sheep ranch.

After disputes between Roussel and Dutrou-Bornier over his land purchases, or more aptly, swindling, and even his kidnapping of young girls for pleasure, things came to a head with a series of violent strikes, including Dutrou-Bornier's torching of the mission's houses and other buildings. Dutrou-Bornier had forcibly married a Rapanui "queen" who bore his children. This, he believed, assured his status as "king," and, moreover, he "owned" much of the island. The violence eventually forced the mission to end its efforts on Rapa Nui in 1871. When the mission quit the island and returned to Mangareva, many Rapanui departed with them. Dutrou-Bornier was now a "king" unfettered by Catholic clergy or other European onlookers. By 1875, his holdings included over 80 percent of the island land, 4,000 sheep, 70 cattle, 20 horses, 300 pigs, and many fowl.[52] He was killed in a Rapanui ambush on the island in 1876, but his devious plot lived on in the form of the removal of islanders from traditional land and the redefinition of the island as devoted to farming practices.

The commercial potential of the island brought interest from the mainland, and Chile annexed Rapa Nui in 1888. Chile's government had strategic, economic, and political interests in acquiring the island. In 1896 Valparaiso merchant Enrique Merlet purchased and leased nearly the entire island in order to establish it as a ranch. The ranch grew rapidly and by 1898 ruled not only the land, but also the people and their labors. The Rapanui workers were ordered to build stone walls—*pircas* in Chilean Spanish—that delineated ranch lands and prohibited Rapanui trespass. This led to prisonlike confinement when the ranch forced the Rapanui to build a nine-foot stone wall completely enclosing the town of Hanga Roa. Rapanui were forbidden to go beyond the wall, to any other part of the island, unless on sanctioned company business. The Rapanui had now become imprisoned laborers on their own island. Rapanui wages went right back to the ranch when they purchased overpriced foreign goods like tobacco and sugar at the company-run store. Eventually some financial problems led the British trading firm, Williamson, Balfour & Company, to buy out controlling shares from Merlet. Business changed, but nothing much else did for the next fifty years of company rule on Rapa Nui.

Overall, the sheep ranch brought possibly the greatest ecological impacts to the island since the arrival of the colonists in around 1200. The grazing of many thousands of sheep for more than sixty years resulted in the extinction of many native plants and caused substantial erosion across the island's steeper slopes. The herds of sheep and cattle that once dominated the landscape today have shaped much of the island's visible surface.

The native population began to rebound from the 1877 low of

just 110. A census taken by the company manager in 1882 comprised a list of every man, woman, and child on the island. The total number of Rapanui was 155, with 68 men, 43 women, 17 boys, and 27 girls under fifteen years of age.[53] Still, there was concern that the islanders might perish entirely. Captain Wilhelm Geiseler, who commanded the German gunboat *Hyäne,* called into Rapa Nui for nearly a week in 1882, remarked that "it is to be expected that if the causes for the decline of population continue, in a short time the last Rapanui man will have lived on his native island."[54]

European visitors around the turn of the twentieth century provide some insight into the life on the island during this time. In December 1886, an American team arrived on the USS *Mohican* with paymaster William Thomson and the ship's surgeon, George Cooke. Thomson conducted a broad survey of the *moai, ahu,* ancient settlements, caves, petroglyphs, and paintings. He and his crew studied the ceremonial village of Orongo in detail, excavated there, and removed painted birdman panels. Covering a significant part of the island, Thomson recorded 555 *moai* and 113 *ahu;* such an extensive inventory would not be replicated until our most recent surveys. He also noted other important details, such as the fact that the island now had 18,000 sheep, 600 cattle, and "a few tough little horses have been introduced from the island breed of Tahiti."[55] He noted that birds, other than chickens, were altogether absent. Thomson offered this observation about an environment ravished by sheep: "In other parts of the island may be seen, in places in considerable numbers, a hardwood tree, more properly bush or brush, called by the natives *toromiro.* These must have flourished well at one time, but are now all, or nearly all, dead and decaying by reason of being stripped of their bark by the flocks of sheep which roam at will all over the island. None of the trees are, perhaps, over 10 feet in height, nor their trunks no more than 2 or 3 inches in diameter."[56] The *toromiro,* a woody shrub, was the last known native plant to have grown in abundance on Rapa Nui. Reports indicate that the last plants grew naturally on the island in the early twentieth century. The sheep appear to have finished them off, likely along with other native plants that might have still grown on the island.

Thomson also provided a vivid image of life on Rapa Nui in 1886 as a consequence of the sheep ranch. He described the locals as living in rough houses constructed from the wood salvaged from the cargo of shipwrecks. These houses, furnished largely by mats for sleeping, had dirt floors and were "vermin infested."[57] Introduced insects, in particular, seemed to have infested the island.

> From the earliest dawn of day to the close of the short twilight, hordes of flies annoyed us; it made no difference whether we skirted the cliffs to windward, climbed the breeze-swept hills, or burrowed in the musty caves and tombs, swarms of flies met us, prepared to dispute every foot of ground. Whatever may have been the parent stock of the Polynesians, we came to the unanimous conclusion that we had discovered here the lineal descendants of the flies that composed the Egyptian plague, and can testify that they have not degenerated in the lapse of time.
>
> Fleas occasioned us more annoyance than the flies, because this industrious little insect was untiring in its attentions by day and night. They were found in numbers in all the camping places, and we seemed to get a fresh supply every time a halt was called.
>
> There are fifteen or twenty mangy dogs of a mongrel breed on the island whose hides were literally alive with jumping insects.[58]

When Katherine Routledge and her husband arrived on the island in 1914, the Williamson, Balfour's sheep ranch was in full operation and the company controlled just about everything on the island. The abuses of the Rapanui by the company and tensions they engendered seemed to have reached a boiling point just a few months into the Routledges' stay. Routledge tells it this way:

> On June 30th, while we were still at the Manager's, a curious development began which turned the history of the next five

> weeks into a Gilbertian opera—a play, however, with an undercurrent of reality which made the time the most anxious in the story of the Expedition. On that date a semi-crippled old woman, named Angata, came up to the Manager's house accompanied by two men, and informed him that she had had a dream from God, according to which M. Merlet, the chairman of the Company, was "no more," and the island belonged to the Kanakas [term used then for native Rapanui], who were to take the cattle and have a feast the following day. Our party also was to be laid under contribution, which, it later transpired, was to take the form of my clothes. Later in the day . . . [a] declaration of war was formally handed to Mr. Edmunds, written in Spanish as spoken on the island.[59]

Angata was a pious Catholic who had worked as an assistant at the church for many years. Her son had penned the letter "declaring war" in both Spanish and Rapanui. The next day, Angata blessed her son and the others who had taken to their horses, rosaries in hand for the protection of God, and rode through the imprisoning wall around the village of Hanga Roa to begin rustling cattle. That night they had their feast of ten slaughtered cattle at the Hanga Roa church.[60] Angata had more dreams, and issued more orders for violence. Edmunds demanded armed assistance from the Routledges and other members of the expedition. More Rapanui engaged in rustling of sheep and cattle, also killing many. Angata made gestures to Katherine for her to side with the Rapanui. Katherine, for her part, tried to dissuade Angata and her followers from more trouble with the company. But Katherine couldn't stop the "word of God" as Angata saw it, and an early Rapanui independence movement to boot. Fearing for their safety, indeed their lives if things escalated, Katherine and her husband, Scoresby, moved to a tent camp at the base of the statue quarry at Rano Raraku, some ten miles from the tensions in Mataveri and Hanga Roa. The Rapanui rebellion would be squashed not by the company but by the Chilean navy, even though the latter took sides with the Rapanui—against the company—in the course of events.

Routledge also noted the lingering effects of diseases that still plagued the island. Leprosy set upon the residents and a number of the inflicted were confined to a leper colony just outside town. She wrote, "unfortunately, some of the old men who knew the most were confined to the leper settlement some three miles north of Hanga Roa. . . . But how could one allow the last vestige of knowledge in Easter Island to die out without an effort? So I went, disinfected my clothes on return, studied, must it be confessed, my fingers and toes, and hoped for the best."[61]

Rapanui remained indentured workers for the sheep ranch through the first half of the twentieth century. Only in 1953 did the era of sheep ranching and the dominance of the Williamson, Balfour company end when the Chilean government refused to renew the company's permit and sheep operations ceased. However, the population remained governed by the Chilean navy, which maintained a permanent base on the island.

When Heyerdahl arrived on the island in 1955 it was still connected to the rest of the world only by the annual visits of a naval ship. Only recently rid of the thousands of sheep that once roamed the landscape, the island was a dusty place largely devoid of trees—the majority of the coconut palm and eucalyptus trees were planted after Heyerdahl's visit. His arrival was an extraordinary event for the islanders: nearly the entire population gathered at the shore to meet him as he landed.

Due in part to the archaeological research that brought the attention of scientists around the world to the island, Rapa Nui has slowly become integrated into the world economy. The U.S. Air Force reconstructed a rough airstrip in the 1960s to provide a base for spying on French nuclear testing. This runway was greatly extended in the 1980s by NASA so it could serve as an emergency landing strip for the space shuttle. With the airstrip, tourism has increased steadily over the years and has provided economic opportunities for the islanders. The town of Hanga Roa has grown from just a few houses at the turn of the century to a population center with shops, hotels, and restaurants. The native Rapanui population size has grown as well. In the 2002 census, the island's population was recorded as 3,791—2,275 of whom are Rapanui, indicating that the population has finally returned to something close to its pre-collapse size.

Over time, the Rapanui have begun to take over the island's governance. In 1965, Alfonso Rapu, the elder brother of our collaborator Sergio Rapu, led a group of residents in an insurrection that forced the Chilean government to return land to the Rapanui. He was later elected mayor of the island. This insurrection led to full Chilean citizenship for the islanders in 1966. And in 1984, Sergio became the first appointed native governor and led the island through 1990.

Thus, despite the long history of disease, population collapse, external rule, and enslavement, the Rapanui have held on and thrived. A swelling population spurred by a booming Chilean economy has brought prosperity to the island in the form of growing tourism. In fact, the irony is that this success now poses its own threat.

The demands of global tourism are straining the island's modest infrastructure. The island's groundwater source, for example, is in imminent danger of being contaminated by sewage, and a communal landfill is bursting at the seams. Jets packed with freight cannot keep enough food on the grocery store shelves—many items run out as islanders await the next flight. Today even a minor disruption in flights would spell crisis, even chaos. Sergio Rapu worries that "if we don't solve these problems of development versus conservation, we may face the island being totally spoiled, turned into a shantytown rather than a beautiful, open-air museum, simply because we didn't learn to take care of it in time."[62]

The island is facing as daunting a set of challenges as it ever has, and it is our hope that the revised account we have offered of the ingenuity with which the Rapanui crafted a life of such delicate ecological balance for so long provides reason for hope that they will overcome the current challenges as well.

CHAPTER 10

Conclusion

> Finding the occasional straw of truth awash in a great ocean of confusion and bamboozle requires vigilance, dedication, and courage. But if we don't practice these tough habits of thought, we cannot hope to solve the truly serious problems that face us—and we risk becoming a nation of suckers, a world of suckers, up for grabs by the next charlatan who saunters along.
>
> —Carl Sagan, *The Demon-Haunted World: Science as a Candle in the Dark,* 1995

> The past is never dead. In fact, it's not even past.
>
> —William Faulkner, *Requiem for a Nun,* 1951

We never expected that our original visit to Rapa Nui would start us on such a journey of discovery. Given the amount of attention the island has received for more than a century and the degree to which its history is held up as the poster child of ecological catastrophes, we fully expected that our work was going to fill in some details, perhaps provide some evidence about the early subsistence and settlement practices in what had every indication of being a well-established prehistory. Maybe, if we were lucky, our excavations in the sands of Anakena might uncover fishhooks with shapes linking early colonists to their homeland in Polynesia. We might add some details to the well-known prehistory, but anything more than that seemed improbable at best.

Yet, as we probed the archaeological record and absorbed others' recent findings, we found that, one by one, the "mysteries" of Rapa Nui became remarkably easy to explain.

The island's history could have been very different. Other islands of East Polynesia were settled at about the same time as Rapa Nui, and their stories provide a window into what might have come to pass. Rapa Iti, in the Austral Islands, is one such. At a similar latitude as Rapa Nui (about 27 degrees south of the equator),[1] Rapa Iti has a similar climate, but it is only 23 percent the size of Rapa Nui, and it is not as isolated from neighboring islands. Rapa Iti enjoys greater rainfall, which supports permanent streams that provide water for irrigating highly productive taro fields. With food productivity therefore relatively high and reliable, rather than living in dispersed settlements as the Rapanui did, and reducing the variance in food supplies through communal activities, those on Rapa Iti ended up concentrated around agricultural fields in fiercely defended territories. The remains of many hilltop fortresses[2] indicate that they turned to intense warfare.

A different fate befell Pitcairn Island. Colonized around AD 1200 and located in a similar southerly latitude, 25 degrees south, Pitcairn is also small, only 29 percent the size of Rapa Nui, and unlike Rapa Iti, it does not appear to have been suitable for wet taro cultivation. The prehistoric population participated in some statue and platform construction, but the island's productivity did not appear to have been able to sustain even dispersed communities. While more investigations are required to determine the exact patterns of ancient settlement on the island and the ecological constraints the islanders coped with, its fate is clear; Pitcairn Island was eventually entirely abandoned. The first Europeans arriving on the island in 1790 found it to be uninhabited.

The disruption of the delicate balance of life on Rapa Nui that followed European contact is familiar in history. It occurred in many parts of the world when once-isolated populations were discovered by European explorers. The great irony is that, having survived that trauma, the Rapanui are now battling the dangers of being connected to the global economy. Overall, the people who live on Rapa Nui today are increasingly reliant on resources that come from the mainland of Chile, and given a swelling population, spurred by a booming Chilean economy and rapidly increasing tourism to the island, the island's modest infrastructure is being greatly strained. Tourism has increased from about 12,000 visitors in 2000 to more than 70,000 in 2010, and estimates by Chile's Ministry of Public Works predict that this number will rise to 200,000 visitors each year by 2020.

Rapa Nui is now facing a new dilemma. While populations living and visiting the island increase and demands grow, the degree to which life on the island depends on the mainland also grows. Rather than maintaining a subsistence level of food supplies, capable of tolerating periodic shortfalls, as was achieved for so many years, the island's population now relies almost exclusively on its link to the larger world economy, and while it has never been more connected to the rest of the world, it has never been at worse risk.[3]

As we have conducted our work on the island, we have often speculated about its future. We initially believed that the breakneck speed of growth and development made an eventual collapse—held off for so long—almost inevitable. Upon reflection, however, we began to wonder whether the contrast between the past and the future is really so stark.

While the island and its occupants will undoubtedly experience continuing change in the coming years, the future may not be catastrophic. When we consider the physical, geographical, and environmental constraints that shaped the island's cultural evolution, we can see that actually little has fundamentally changed since the first Polynesian colonists arrived some 800 years ago. Even with jet travel and contemporary communication systems, the island is still remote and the local options are still limited. Despite modern cultivation practices, the soil remains nutrient poor, and seafood is not especially abundant. Choices on Rapa Nui must still always be made with an eye to the island's finite land and food sources. And the people still make those choices knowing they are family matters: everyone on the island can trace family connections to almost everyone else. They of course have their quarrels, but greater direct conflicts are assiduously avoided.

Thus, we have what we might call the Rapa Nui Effect continuing to this day, still shaping the island's destiny, and we've come to believe that the Rapa Nui Effect makes it more than likely that the Rapanui will persist. The historical record suggests that, in general, populations forced to confront challenging local conditions with the wisdom of local knowledge persist and even thrive. Indeed, we believe that we all would do well to study the success of Rapa Nui when we consider the global challenges we now face. Given the worries about the environment, resource scarcity, increasing conflict, and growing populations, the lessons learned by the prehistoric Easter Islanders have never had greater significance.

More than ever, we face a series of choices that require us to balance considerations of short-term benefits and long-term stability. Estimates are that our populations will continue to grow and our resources will become more limited, making these choices perhaps the most important issue of our times. We hope that the history of Rapa Nui can be an inspiring vision of human ingenuity in facing such challenges and of human resilience.

APPENDIX 1

Environmental Constraints

Unlike the deep valleys, steep mountains, jagged ridgelines, streams, and waterfalls typical of the high volcanic islands of Tahiti or the Marquesas, Rapa Nui is a modest landscape. The island was born less than a million years ago when the coalescing eruptions of three seafloor hotspot volcanoes reached the surface first with Poike, joined by Rano Kau, a caldera of intermediate age, and finally by Maunga Terevaka, with its daughter cinder cones along active rift zones. The youngest and largest volcano, Maunga Terevaka is dotted with numerous cinder cones—volcanic craters—that exploded out of the island's broad slopes over recent geologic time. Polynesian colonists must have been pleased to find abundant high-quality obsidian on cinder cones as well as on the offshore islets. Other stone resources, such as fine-grained basalt for fashioning adzes, were soon discovered and utilized. Polynesians recognized the malleable volcanic tuffs of red and gray, such as those found in abundance at Puna Pau and Rano Raraku. These kinds of stone were familiar raw materials for carving images to show respect for the ancestors.

The volcanic summit of the island at Maunga Terevaka reaches just 1,676 feet, an elevation inducing little, if any, orographic[1] effect to bring rainfall like those of the greater peaks of other Pacific islands. A higher peak snaring the winds and moisture-laden clouds would have been a welcome blessing to Rapa Nui, especially since the island has no permanent streams. It probably never did. Water is found in three lakes formed in volcanic cones: Rano Kau, Rano Raraku, and Rano Aroi, near the summit of Terevaka. Water also flows from a number of small springs, in caves, and from wells dug deep into the water table. Unlike the accessible flow of stream water for irrigation of taro, Rapa Nui's lakes essentially trap water. Elsewhere in Polynesia streamflow was diverted into ditches that fed pond fields of irrigated taro. Such cultivation was not only highly productive, but wonderfully predictable. Permanent streams meant a regular supply of water for irrigation, making taro *four times* more productive than taro under nonirrigated conditions. And irrigated taro fields are eight times as productive as yams. As archaeologist David Addison has pointed out, irrigated taro, with its highly productive and predictable yields, would have been a critical crop supporting initial colonization.[2] Yet on Rapa Nui, irrigation and the familiar pond field bounty of taro was simply not an option.

As options for establishing the mainstay of irrigated taro were dashed, the colonists of Rapa Nui must have watched in dismay as critical crops such as breadfruit simply died. Polynesians brought tropical cultigens to the subtropics, where the failure of some, such as breadfruit and coconut, was imminent. With the loss of breadfruit went a major staple. Worse, the loss of breadfruit meant the loss of a crop that could be prepared into a paste and fermented for long-term storage in subterranean pits. In the Marquesas Islands, for example, storage of this tasty yogurtlike, starchy breadfruit paste (*ka'aku*) provided an insurance policy against food shortages for literally decades. As the ancient Rapanui would discover, such a food storage strategy would have served them well. But it wasn't to be—not on this island.

Table A1.1. Major Polynesian Cultigens and Their Range of Optimal Growth in Latitude

Plant (taxon)	**Range of growth (latitude)**[a]	**Grows/Fruits on Rapa Nui?**
Coconut (*Cocos nucifera*)	N26 to S10 degrees	No[b]
Kava (*Piper methysticum*)	N25 to S20 degrees	No
Breadfruit (*Artocarpus altilis*)	N23 to S17 degrees	No

Plant (taxon)	Range of growth (latitude)	Grows/Fruits on Rapa Nui?
Taro (*Colocasia esulenta*)	N35 to S18 degrees	Yes
Yam (*Dioscorea alata*)	N23 to S20 degrees	Yes
Sweet Potato (*Ipomoea batatas*)	N40 to S32 degrees	Yes
Banana (*Musa sapientum*)	N31 to S31 degrees	Yes
Sugarcane (*Saccharum officinarum*)	N35 to S35 degrees	Yes

[a] Modern estimates for optimal agricultural potential.
[b] Some important food crops, such as coconut and breadfruit, would not have survived on Rapa Nui if they had been introduced. With recent climate change, Rapa Nui is now becoming warm enough for coconuts to fruit. In prehistoric times, however, this was not the case, and the coconut was absent.

Source: Data from the Food and Agriculture Organization (FAO) of the United Nations.

There would have to be other solutions to life on this humble land. With perhaps as few as 30, or as many as 100, the newly arrived colonists would have initially enjoyed ample wild resources. But they had their work cut out for them in establishing crops. They also did what small, colonizing populations need to do, and always seem to do: they reproduced. Given the opportunities and the competition that would soon emerge, the natural thing—even the necessary thing—would be to produce large families. A decision to do otherwise, to have just one or two children, would mean your neighbors or more distant kin would soon be taking over—land, resources, cooperative labor, control over decisions, and so on. In this new colony, even with its limitations, there would be no immediate rewards for having smaller families.

Real-world, detailed examples of just what happens with small colonizing groups when they reach a new island or territory are actually hard to come by. But we have one from Pitcairn. The mutineers from the *Bounty* who settled with their Tahitian partners on this tiny resource-poor island in 1767 reproduced at a phenomenally rapid pace. In eighty-nine years the founding population of

15 men and 12 women, and an infant girl, had grown to a total of 193. This works out to be as high as 3.7 percent annual growth. This roughly doubling of the population about every twenty years could not be sustained indefinitely, but characterizes the exponential growth biologists expect to see in colonizing groups.

The *Bounty* mutineers offer an historic glimpse of what population growth must have been like on Rapa Nui in the first decades following colonization. In fact, human demographers have demonstrated in mathematical simulations that small founding populations, especially those isolated from other potential mates, must grow rapidly or face likely extinction.[3] Extinction sounds harsh, but it simply describes what happens when there are too few individuals of the right age and sex to perpetuate a population over time. We can surmise, then, that the small founding population would, by nature and necessity, have grown quickly. People had large families and began to disperse themselves over the island's landscape in a relatively short time. The growth rates meant that in a century or less, the populations could have grown to more than a thousand. As population reached a sustainable threshold, as we have seen in so many real-world cases, growth slowed and numbers held at a "demographic carrying capacity," sometimes also considered an "environmental carrying capacity," depending on food production technology and interactions with other populations, among other things.

Rapa Nui was not like other islands. Some crops simply died, summarily removed from the agricultural roster at this subtropical latitude. Lacking permanent streams, there was no possibility for the higher productivity of irrigated taro, an agricultural strategy that supported Polynesian populations elsewhere. A growing population faced with these limitations meant that the regularity of rainfall gained genuine importance. The challenges and difficulties of agriculture soon became painfully evident. Rainfall was moderate, but worse, unpredictable.[4] There were very wet months, wet years, dry months, or even years of low rainfall. Climatologists objectively reckon precipitation in terms of surpluses and deficits. For the ancient Rapanui farmers, the unpredictable rainfall was potentially the difference between prosperity and destitution. We can think of dealing with uncertain rainfall as analogous to getting by in today's world with a periodic paycheck; you wouldn't know when you would receive it, and then add the frustration of not knowing how much you would be paid. You might get a lot, then one week later get even more; a month later nothing. The best you could estimate would be that some seasons tend to be better than others. Add to this analogy having no other job options, no chance of moving somewhere else for the hopes of greener grass.

Today rainfall on Rapa Nui averages just over thirty inches a year. On a month-to-month basis, more rain tends to come during the Southern Hemisphere winter (June–August). Yet averages can be deceptive and patterns irregular. In some years the winter can be dry, while in others the summer can be wet. A "normal" year in average rainfall can also mask unevenness where several inches fell in one month, with little if any rainfall in others. We must also consider rainfall in terms of how much water is available to plants and how much water they need. Evapotranspiration is a measure of evaporation and plant transpiration from the earth's surface into the atmosphere. The evaporation part of the equation accounts for movement of water into the air from sources such as the soil, bodies of water, and vegetation. Transpiration, on the other hand, is the movement of water within plants and the subsequent loss of water as vapor from its leaves. We are basically talking about the effective amount of drying and its variability under different conditions. Rainfall matters in relation to the amount of evapotranspiration, among factors such as drainage and runoff.

Table A1.2. Rainfall, Temperature, and Evapotranspiration Statistics for Rapa Nui, 1958–97

Month	Evapotranspiration (in.)	High Temp (F)	Mean Temp (F)	Low Temp (F)	Rainfall Average (in.)	Rainfall Standard Deviation (in.)
January	6.7	79.8	73.4	67.1	2.9	2.6
February	5.4	81.0	74.1	67.1	2.9	2.6

APPENDIX 1

Month	Evapo-transpi-ration (in.)	High Temp (F)	Mean Temp (F)	Low Temp (F)	Rainfall Average (in.)	Rainfall Standard Devia-tion (in.)
March	4.6	80.2	73.7	67.2	2.9	2.6
April	3.3	77.5	70.8	64.3	2.8	2.5
May	2.8	75.3	69.0	62.3	2.7	2.5
June	2.3	71.4	65.9	60.0	2.6	2.4
July	2.4	70.5	64.6	59.0	2.5	2.3
August	2.6	70.3	64.5	58.8	2.5	2.3
September	3.4	70.9	65.0	58.5	2.6	2.3
October	4.2	72.8	66.0	58.7	2.6	2.3
November	4.6	74.6	68.3	61.7	2.7	2.4
December	5.6	78.0	70.9	64.4	2.8	2.5
Total	47.9				32.5	
Average	4.0	75.2	68.8	62.4	2.7	2.5
Standard Deviation	1.4	4.0	3.7	3.5	0.1	0.1

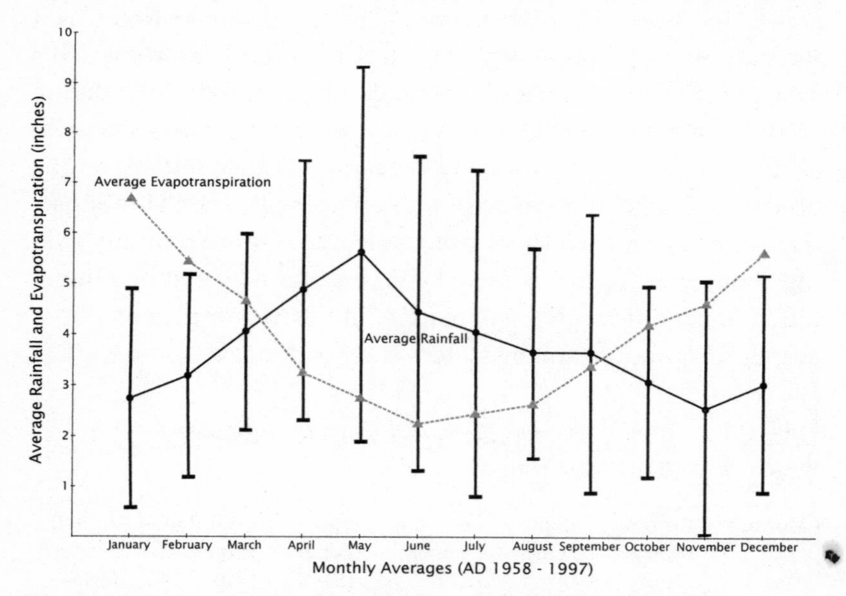

Figure A1.1. Average monthly rainfall, its variability (standard deviation), and average evapotranspiration on Rapa Nui from 1958 to 1997. Note that under average conditions, evapotranspiration exceeds rainfall six months of the year.

When evapotranspiration exceeds rainfall, we run into the technical definition of a desert environment, or drought conditions when the deficit surpasses the average for some extended period. Albeit variable and unpredictable, under average conditions Rapa Nui faces desert conditions over about six months of the year. As depicted in Figure A1.1, p. 186, and detailed in the table, dry, desert-like conditions are met when evapotranspiration falls below average rainfall. Six months is a long time for a plant. In Figure A1.2, we compare annual plant requirements[5] with rainfall and evapotranspiration, illustrating the critical conditions ancient Rapanui farmers must have faced from year to year. Plotting the requirements of these major Polynesian cultigens against rainfall and evapotranspiration also reveals the crops that could best tolerate the difficult conditions on Rapa Nui. For example, sweet potatoes, bananas, and yams could tolerate average conditions. Taro and sugarcane had higher demands and would fall outside their minimal requirements over at least several of the years plotted with these modern data.[6]

If water posed problems for cultivation, it was only part of the story for the ancient islanders. Wind brought additional challenges. Average conditions bring winds blowing between about 5 to 35 miles per hour. Unfortunately, the winds are also frequently strong, sometimes blowing for days on end. Plotting winds recorded at the airport on Rapa Nui from 1958 to 2005 illustrates

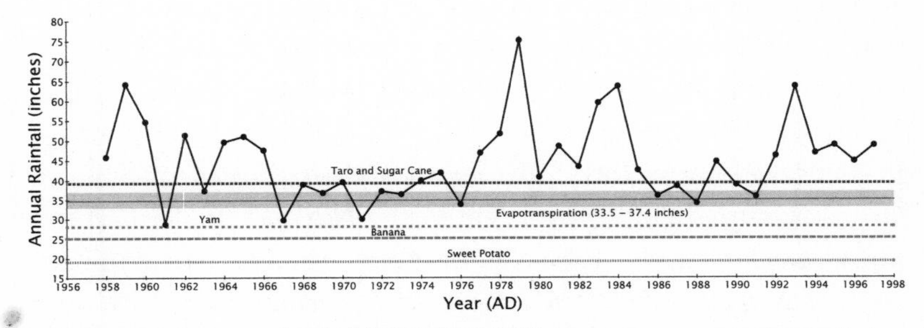

Figure A1.2. Annual rainfall (1958–1997), average evapotranspiration, and minimal moisture requirements for the major Polynesian cultigens on Rapa Nui. Data on plant growth and agricultural potential are from the Food and Agriculture Organization of the United Nations (FAO).

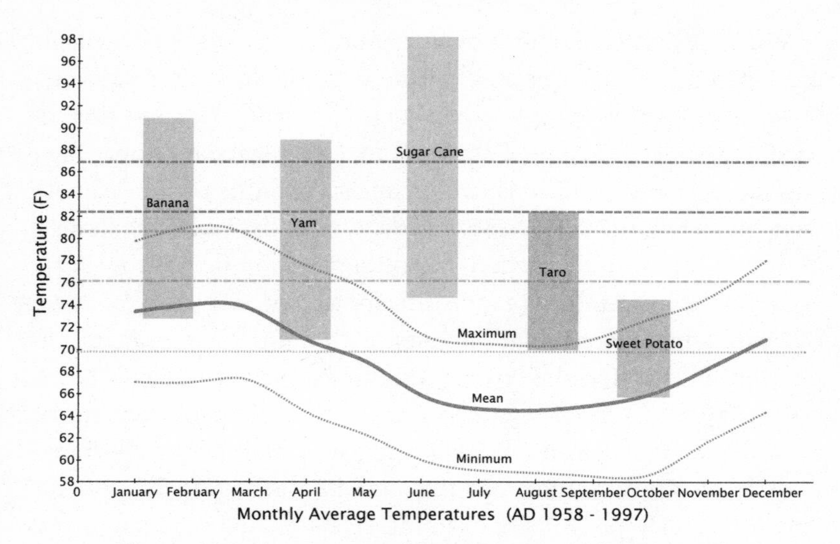

Figure A1.3. Average monthly temperatures (1958–1997) and temperature requirements for the major Polynesian cultigens on Rapa Nui. Data on plant growth and agricultural potential are from the Food and Agriculture Organization of the United Nations (FAO).

what happens. In just the forty-seven years recorded, hurricane-force winds (greater than 74 miles per hour) ripped through the island at least ten times. Strong sustained winds would soon become laden with the burning effects of salt off the ocean. When the winds blow for a day or more, you can taste the salt in the air and your skin begins to feel tight, dry, and even crusted with salt. The salt spray has encrusted and destroyed the lenses of more than a few of our sunglasses. Under these kinds of conditions it doesn't take long for the tender leaves of banana trees to be burnt, shriveled, curled, and brown. Often the bananas don't survive a few days of strong wind. Taro withers and dies with damaging salt spray as well. The frequent strong winds certainly took their toll on ancient crops on Rapa Nui.

Many of the conditions for agriculture changed as the island was transformed with human colonization. As forest loss proceeded over several centuries, removal of vegetation meant sur-

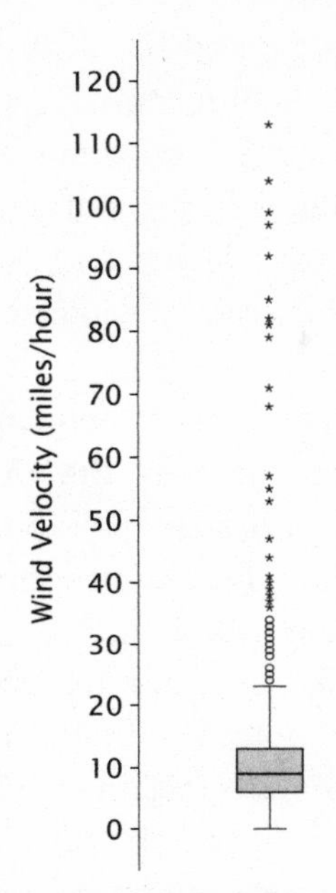

Figure A1.4. Average and stronger sustained winds ("outliers") recorded at the Mataveri Airport on Rapa Nui between 1958 and 2005. Over this 47-year recording period, hurricane-force winds (greater than 74 miles per hour) occurred at least ten times, or roughly once every five years.

face wind would increase in velocity and its effects, as well as leading to increased evapotranspiration. Gradually deforestation exacerbated some problems for agriculture, even as it opened more land for cultivation.

The subtropical climate, strong winds, and the surpluses and deficits of uneven rainfall were not the only problems in growing the crops essential to the island's subsistence. Soil posed another tremendous challenge for the ancient Rapanui. The island's relatively young geologic age explains shallow soils and extensive areas of volcanic rock outcrops at the surface. The soils are also often accurately described as excessively well drained. They don't hold moisture for long, making them less supportive for plant growth. But the worst aspect is that this young island is old enough for soils formed in volcanic ash and subjected to weathering to have lost vital nutrients, especially phosphorus and potassium.[7] We see some of the same problems in the Hawaiian Islands. Our colleague David Burney, who works diligently to grow endangered native plants and restore their ecosystems, complains that the weathered volcanic soils in Hawaii "have the fertility of a brick."

Field research on Rapa Nui by Belgian soil scientist Geertrui Louwagie and his colleagues has shown that the island's soils are just moderately to marginally suitable for growing crops. On some areas of the island where weathering has taken even more of a toll, the soils are described simply as "not suitable" for cultivation. Rapa Nui's climate, moisture, and limited soil nutrients would leave the sweet potato with the most reliable potential. In the less than optimal conditions of Rapa Nui, sweet potatoes could normally withstand fluctuations in moisture supply and still produce two crops each year. The size and quality of the tubers depended on the soil fertility. Louwagie and his colleagues also document that weathering and erosion down to the less fertile soils occurred long ago, before deforestation, before humans ever reached the island.[8] The loss of any truly fertile soils that once existed was *not* a consequence of deforestation, as some have assumed,[9] but rather a natural process unaided by humans that happened eons ago.

Rapa Nui had its difficulties. It wasn't exactly paradise, but neither was it hell. The island had its potential tempered by its limitations. We think this in-between status may be the key to understanding so much about what happened in Rapa Nui's human story. Despite the challenges, Polynesians made a living on Rapa Nui for more than five hundred years before Europeans sailed over the horizon to this most remote corner of the world. When the visitors did come, things changed forever.

APPENDIX 2

Lithic Mulching and *Manavai*

On Rapa Nui, soils were made adequate for cultivation through what is known as "lithic mulch."[1] This is a term that generally describes the addition of rocks to areas of cultivation. Lithic mulching, sometimes known as "plaggen soils," is relatively common around the world and examples can be found in both prehistoric and historic contexts.[2]

To examine how stone features affect temperature, one of our graduate students, Alex Morrison, made temperature measurements over two weeks for a location just inside and outside a *manavai* (Figure A2.1, p. 192). While the average temperatures were similar, the inside of the *manavai* experienced smaller swings in temperature relative to the temperatures just outside the *manavai*. This effect is due to the wind shelter as well as heat absorbed and released in the night by surface rocks. This experiment also shed light on why *manavai* are often located adjacent to cave entrances. Temperature measurements for a *manavai* that is located immediately outside a cave have even smaller diurnal variability than a *manavai* not located near caves.

The suppression of temperature variability appears to be a general result of rock features used in cultivation. Even just thin layers of cobbles on the surface have the same effect. Figure A2.2 (p. 193) shows measurements made by Alex Morrison for a temperature variability measured over a week inside and outside a stretch of lithic mulch. Others have also noted this property of lithic mulch. Studies conducted by archaeologist Chris Stevenson

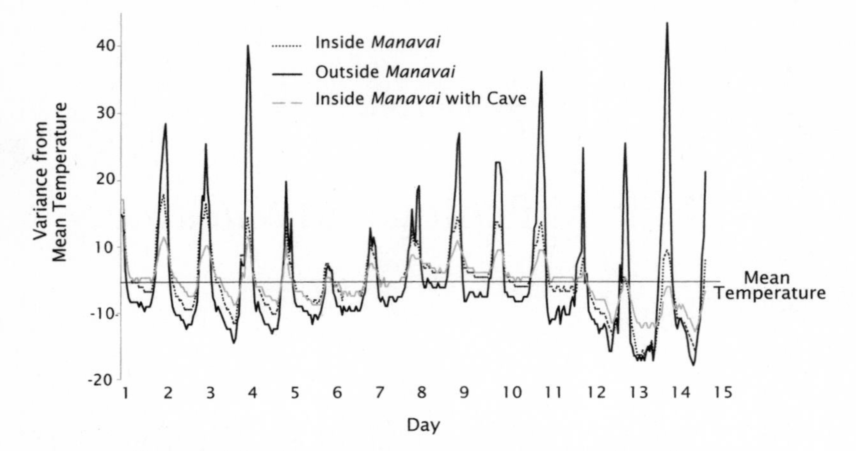

Figure A2.1. Temperature measurements made in three locations over two weeks. Temperature is measured relative to deviations from the overall mean temperature of the period. During the daytime, temperatures rise to a peak, then they fall to lows at night. The lines show temperature change on the outside surface, inside a *manavai* structure, as well as inside a *manavai* with an adjacent cave. Note how the range of ups and downs is smaller for the temperatures inside the *manavai,* particularly for those located adjacent to a cave.

and colleagues for lithic mulch areas consisting of rock veneers and boulder gardens have shown similar kinds of patterns.[3] The addition of a layer of rocks to a garden mediates temperature swings, producing a more stable environment for plant growth.

Lithic mulch features on Rapa Nui come in a variety of forms.[4] They range from a single continuous layer of small 2–8 inch diameter rocks (often called lithic pavements or veneers) to rock gardens consisting of one or more layers of densely concentrated rocks that are roughly 2–5 inches in diameter, to boulder gardens that are formed from larger, 12–32 inch basalt chunks, to deep piles of rock with multiple depressions (known as *poe poe*). While they are often given different names, these cultivation features—from *manavai* and simple scatters of rocks—form a continuum of habitats for plants. At one end of this habitat spectrum are constructed stone enclosures with deep central pits and tall walls for

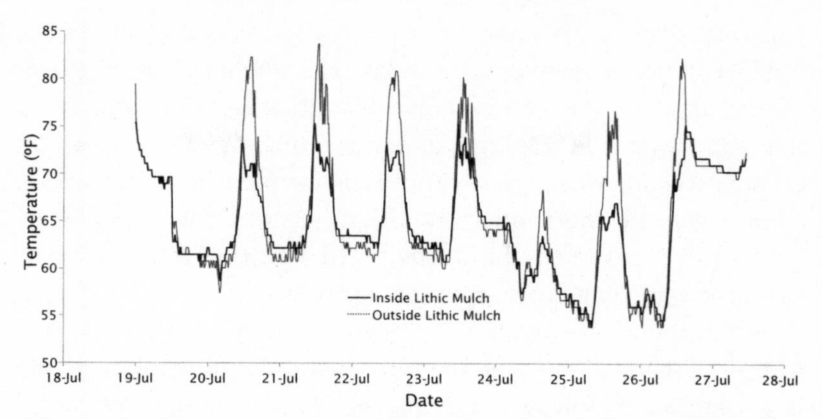

Figure A2.2. Temperature measurements made inside and outside lithic mulch areas on the south coast of Rapa Nui. Lithic mulch, like the *manavai,* reduces amplitude in diurnal temperatures.

protecting tall plants such as bananas and sugarcane. At the other end of the spectrum are lithic veneers with enough space between to shelter the sprouting leaves of sweet potato. In the middle we find piles of rocks of various configuration and height that range from tall layers to piles with depressions for plants with larger leaves, such as taro. Since winds in the region around Rapa Nui have no single predominate direction, the shapes of all of these circular and amorphous structures are well suited for providing shelter as they work equally in all directions.

The placement of rocks, particularly broken, smaller ones, also serves another essential function: it increases the productivity of the soil by exposing fresh, unweathered surfaces and thus mineral nutrients that are within the rock. Often the rocks are placed not only on the surface but also directly in the subsurface to directly introduce new sources of minerals into the soil. This form of stone mulching is a hidden yet vital part of the subsistence practices on prehistoric Rapa Nui.

Studies conducted by soil scientists Geertrui Louwagie and Roger Langohr demonstrate that the abundance of water was never a major problem for prehistoric cultivators. Instead, soil nutrients are the primary limiting factor for the growth of native cultigens sweet potato, taro, yam, sugarcane, and banana.[5] Using models of crop growth coupled with experiments of crops grown in four areas on Rapa Nui, Louwagie and Langohr demonstrated that only with the addition of lithic mulching were soils rich enough to support *marginal* conditions for plant growth. In fact, comparisons of soils from areas with lithic mulch made only marginal differences in the productivity, though that difference was apparently adequate to support crops, at least most of the time.

While the idea of adding rocks to a garden might seem odd to Euro-American farmers, the introduction of mineral nutrients is a common organic farming practice today—though we tend to associate mineral fertilizers with material that has a mechanically reduced, powdery consistency. In nonindustrial populations, mineral additions to supplement soil can be accomplished by breaking rocks into small pieces. Small broken rocks have exposed fresh surfaces that provide minerals from the inner portions, which are less weathered than the exterior.

Thus the reduction of the size of the rocks is also part of the lithic mulching strategy. In areas without sources of sand, gravel, or pebbles to use as mulch, breaking large, weathered rocks into smaller cobbles can result in material suitable as fertilizer. By breaking down large rocks into small pieces one can maximize the amount of exposed surface area available for mineral leaching. Relative to a single large boulder, many hand-sized rocks of the same total volume have many times the amount of surface area. For example, imagine a cube that is 4 inches on each side. The volume of this cube would be 4 x 4 x 4 = 64 cubic inches and its surface area would be 4 x 4 x 6 = 96 square inches. Now imagine dividing the cube into a series that are just 1 inch on each side. Then each cube would be just 1 x 1 x 1 = 1 cubic inch. To make up the same volume as the original big cube, you would need 64 cubes. The surface area of each 1-inch cube, however, is 6 square inches. This makes a total surface area for all of the 1-inch cubes of 64 x 6 = 864 square inches. That is a ninefold increase in surface area using the same volume.

As the size of the smaller parts get smaller, the relative increase in surface area continues to grow at an exponential rate. This relation explains why industrial mechanisms are typically used to crush mineral rich rocks to powder when they are used for fertilizer. In cases where only manual power is available for breaking rocks, we expect that there is a point of diminishing returns for the energy required to make rocks smaller relative to the surface area reduction. In the case of Rapa Nui, prehistoric populations used slightly harder basalt hand axes (*toki*) to break rocks down to smaller chunks. When the rocks reached roughly hand size, making them smaller would be increasingly difficult without technology for crushing. We now have an explanation for the endless expanses of ankle-twisting, billiard-ball-sized rocks: prehistoric cultivation.

Remarkably, due to the productivity-enhancing properties as well as the moisture benefits gained by the surface rocks, prehistoric farmers were also able to reliably grow their crops and withstand periodic drought-linked water shortage. While extremely labor-intensive and not a recipe for bountiful harvests, lithic mulching was an appropriate and effective means of long-term survival on this isolated volcanic island. Quite to our surprise, we found that one solution to the poor soils of Rapa Nui turns out to be lithic mulching and these "rock gardens." While resulting in a relatively low return of crops, the use of abundant rock resources provides a means of slight additions to the soil adequate enough to reliably produce crops in most conditions.

From a plant nutrient perspective, we can now place *manavai* features in context with lithic mulch. *Manavai* serve as wind protection for large plants but they also provide a stable environment for plant growth. Necessary mineral nutrients can be added to *manavai* as they provide a discrete and cumulative location for dumping household debris, ash, organic material, and other substances that will increase soil nutrients to encourage plant growth. Thus *manavai* are high investment in terms of the continuous work required for a single location, yet they have a reliable and relatively high return for any plant they shelter. In contrast, lithic mulching allows larger areas to be cultivated for sweet potatoes as well as taro and yams but without requiring constant investment of organic mulch. With relatively limited sources of organic matter, it is not possible to sustain large field systems by adding organics. However, lithic-mulched fields can be productive, even if only at marginal levels. In this way, lithic mulching enables large fields to be cultivated to produce a reliable though not particularly bountiful source of food. Each kind of feature plays a role in providing the subsistence base to consistently support populations in a way that minimizes risk due to droughts and short-term conditions.

During our fieldwork, we have had opportunities to assess this model as it pertains to cultivation features. Gabe Wofford, one of our undergraduate students from the University of Hawaii, conducted a study of soil composition in and around six *manavai* features located along the north and northwest coasts.[6] Wofford carefully collected samples along a line that extended across each *manavai*. These sample locations include areas outside each *manavai*, three from the inside and one that was located three hundred feet away. Wofford collected the last sample in order to assess the composition of the background environment relative to that of the area around each *manavai*. Back in the lab at the University of Hawaii, Wofford then analyzed each sediment sample to determine the concentration of potassium, phosphorus, calcium, and magnesium.

His findings demonstrate that these features serve as focus points for adding materials that enhance the productivity of the soil. In Figure A2.3, we show the plan of a *manavai* located along the northwest coast at a location known as Maitaki Te Moa. Sample locations are identified along a transect that runs roughly north to south. Figure A2.4 shows the results of the analysis from eight sample locations.

We found this pattern at all of the *manavai* we examined (Table A2.1). Consistently, the concentrations of essential mineral nutrients are found to be significantly higher inside the *manavai* than outside and the difference is often two or three times as great. *Manavai* are associated with efforts that increase plant productivity by enhancing the concentrations of mineral nutrients.

Table A2.1. Summary of Mineral Concentrations for All Eight Manavai *Studied*

Mineral Nutrient	Minimum	Maximum	Mean Inside ($\bar{X}$)	Mean Outside ($\bar{X}$)
P (μg/g)	46.00	6985	1030.21	313.64
K (μg/g)	24.00	2374	770.11	414.43
Ca (μg/g)	588.00	5672	2326.11	1713.71
Mg (μg/g)	284.00	2364	1039.16	880.29

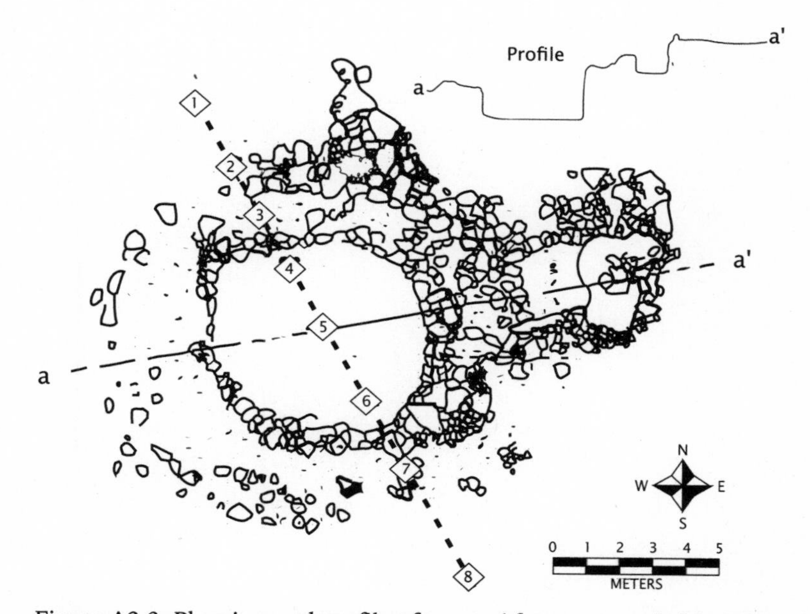

Figure A2.3. Planview and profile of *manavai* feature recorded in 2004. The dotted line shows the transect along which we collected soil samples. Each diamond shows a sample location. The line from *a* to *a'* marks the alignment of the profile in the upper right corner.

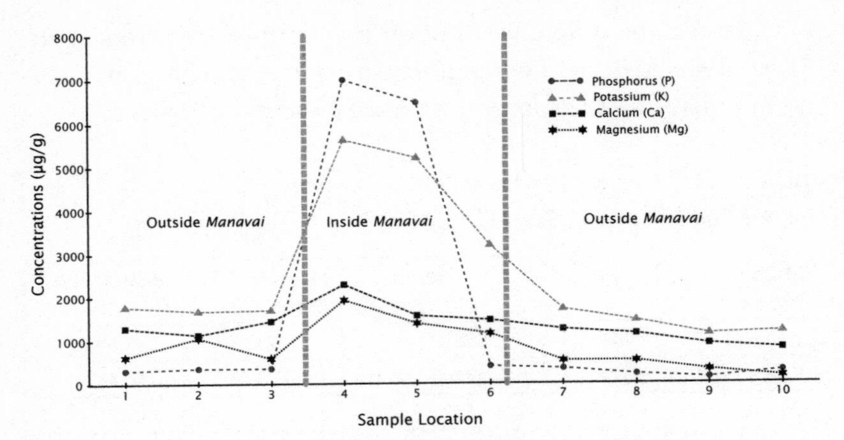

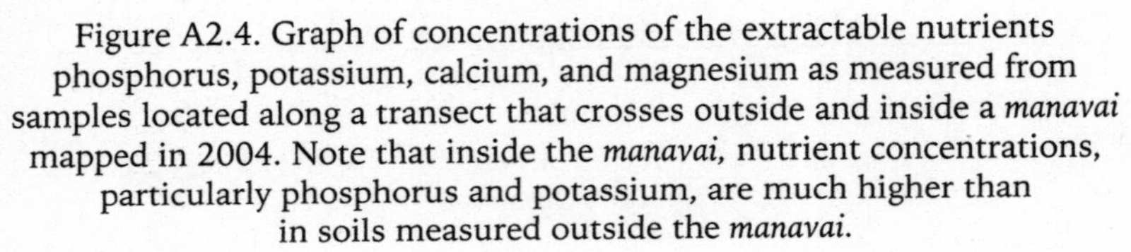
Figure A2.4. Graph of concentrations of the extractable nutrients phosphorus, potassium, calcium, and magnesium as measured from samples located along a transect that crosses outside and inside a *manavai* mapped in 2004. Note that inside the *manavai,* nutrient concentrations, particularly phosphorus and potassium, are much higher than in soils measured outside the *manavai.*

Notes

CHAPTER ONE.
A MOST MYSTERIOUS ISLAND

1. Skottsberg 1956:426.
2. Wilmshurst et al. 2011; for Rapa Nui see Hunt and Lipo 2006, 2008.
3. Scaglion 2005; Storey et al. 2007; Gongora et al. 2008.
4. Ruiz-Tagle 2005:288.
5. La Pérouse 1968:318.
6. Heyerdahl and Ferdon 1961:395.
7. Heyerdahl and Ferdon 1961:33–43.
8. Routledge 1919:211.
9. Heyerdahl 1976, 1989; Heyerdahl and Ferdon 1961.
10. Heyerdahl 1976:29.
11. Heyerdahl 1989:29.
12. Smith 1961b:391; Routledge 1919:280.
13. Diamond 2005:118.
14. At the time many of these were reported in a synthesis by Martinsson-Wallin and Crockford, 2002.
15. Mann et al. 2008.
16. Wilmshurst et al. 2011.

CHAPTER TWO.
MILLIONS OF PALMS

1. Heyerdahl 1961:519; Mulloy 1961:94; Smith 1961a:204.
2. Mulloy and Figueroa 1978:22.
3. Diamond 1995:68.
4. Diamond 2005:109.
5. Bahn and Flenley 1992:214.
6. Ruiz-Tagle 2005:23–24.
7. Ruiz-Tagle 2005:37.
8. Von Saher 1994:99.

9. Behrens 1903:135.
10. Behrens 1903:137.
11. Palmer 1870:168.
12. Dransfield et al. 1984; Flenley et al. 1991.
13. Flenley 1993b.
14. Flenley 2010:22–24.
15. Butler et al. 2004; Butler and Flenley 2001, 2010.
16. Mann et al. 2003, 2008.
17. Orliac 2000, 2003; Orliac and Orliac 1998.
18. Mieth et al. 2002.
19. Mann et al. 2003.
20. Mann et al. 2003.
21. Mieth and Bark 2004:82, 89.
22. Athens et al. 2002:63.
23. Athens 2009.
24. *Sydney Morning Herald,* October 31, 2009.
25. *Sydney Morning Herald,* October 31, 2009.
26. Auld et al. 2010.
27. Huffington Post, April 10, 2010 (updated).
28. Fenchel 1974.
29. At a latitude comparable to Rapa Nui, but with a lower abundance of food, Kure Atoll (28° 24' N) in the northwest Hawaiian Islands supports Polynesian rat densities averaging 45 per acre, with maximum recorded densities reaching 75 (Wirtz 1972). At a density of only 45 rats per acre, Rapa Nui would have supported a rat population over 1.9 million. At 75 per acre, perhaps a reasonable density given the *Jubaea* nut and other forest resources, the rat population of Rapa Nui could have exceeded 3.1 million.
30. Dransfield et al. 1984:750.
31. Flenley et al. 1991:104.
32. Diamond 1985:602.

CHAPTER THREE. RESILIENCE

1. Cook 1777.
2. Thomas and Berghof 2000:322.
3. Orliac and Orliac 1998.
4. Ruiz-Tagle 2005:288.
5. Ladefoged et al. 2003.
6. Caborn 1957.
7. Ruiz-Tagle 2005:37.
8. Fischer 2005.
9. Irwin 1992.
10. Ferdon 1961; McCoy 1976; Vargas Casanova 1998; see Figures 3.2, 3.3.
11. Beechey 1831:41.
12. Mieth and Bork 2003, 2004.

13. La Pérouse 1968:333.
14. Métraux 1940:41.
15. La Pérouse 1968:318.
16. Stevenson and Wozniak 1999.
17. Bork et al. 2004
18. Ladefoged et al. 2005.
19. Louwagie et al. 2006.
20. Vezzoli and Acocella 2009.
21. The age of volcanic eruptions is related to productivity due to weathering. Initially, the soil that is formed from freshly erupted volcanic ash and rock contains abundant nutrient minerals, particularly nitrogen (N), phosphorus (P), and potassium (K) but also calcium (Ca), magnesium (Mg), and sulfur (S). These minerals are essential for plant growth. Phosphorus and nitrogen, for example, are vital parts of photosynthesis, the conversion of solar energy to chemical energy. The addition of fertilizers to modern crops usually is done to provide plants adequate abundances of these minerals to assist in root growth, blooming, and fruit production.

    Generally, when volcanic ash and rock first weathers into material that is suitable for plant life, it contains plenty of N, P, and K. Over time, however, the availability of these minerals for plants declines due to leaching from rainwater and the activity of plants. Consequently, while young volcanic islands are some of the most biologically productive places on earth, islands with old volcanoes can be biological deserts, even with adequate rainfall. Abundant rain can even exacerbate the situation, as the greater the amount of rainfall, the quicker mineral nutrients are flushed from the soil. For this reason, even desert soils can be more productive than volcanic soils simply due to the lack of rain and its leaching effects.

    In the case of Rapa Nui, hundreds of thousands of years have passed since the volcanoes that produced the island were active. Through millennia of rainfall, the soils have been depleted of their nutrients. The island, therefore, has always been a poor place in which to make a living as a farmer. It was poor long before people arrived in AD 1200. It is still a poor place to grow food.
22. Lithic mulch features on Rapa Nui come in a variety of forms. They range from a single continuous layer of small 2–8 inch diameter rocks (often called lithic pavements or veneers) to rock gardens consisting of one or more layers of densely concentrated rocks that are roughly 2–5 inches in diameter, to boulder gardens that are formed from larger, 12–32 inch basalt chunks, to deep piles of rock with multiple depressions (known as *poe poe*). *Manavai* should also be considered part of this continuum. At one end of this spectrum are constructed stone enclosures with deep central pits and tall walls for protecting tall plants such as bananas and sugarcane. At the other end we have lithic veneers with enough space between rocks to shelter the sprouting leaves of sweet potato. In the middle, we find piles of rocks of various configurations and height that range from tall layers to piles with depressions for plants with larger leaves such as taro. Since winds on Rapa Nui are variable, the generally circular shapes of so many of

these structures are well suited for providing shelter as they work equally well in all directions.

23. Reich 1896.
24. Heyerdahl 1989.
25. Mieth and Bork 2003, 2004, 2005.
26. *Manavai* potentially tell us something about at least the magnitude of population. As described earlier, our studies of satellite images indicate that at least 2,500 *manavai* are today found on the island, and this number probably underestimates the maximum present in ancient times, since so much land where *manavai* once existed has seen historic and modern changes. Given the amount of food cultivated in each *manavai,* in conjunction with other forms of subsistence, their numbers are at least consistent with early historic population estimates of around 3,000.

CHAPTER FOUR. THE ANCIENT PATHS OF STONE GIANTS

1. Ruiz-Tagle 2005:284.
2. Ayres and Ayres 1995.
3. Routledge 1919:140.
4. Routledge 1919:194.
5. Routledge 1919:194.
6. Heyerdahl 1989:225.
7. Heyerdahl 1989:225.
8. Love 2001.
9. Allen and Richards 1983; Custer 1986; Failmezger 2001; Fowler 1996, 2002; Kennedy 1998; Mumford and Parcak 2002; Philip et al. 2002; Sever and Wagner 1991; Ur 2003.
10. Lipo and Hunt 2005.

CHAPTER FIVE. THE STATUES THAT WALKED

1. Heyerdahl 1989:226.
2. Heyerdahl 1989:227.
3. MacIntyre 1999:71.
4. "Secrets of Lost Empires: Easter Island," *Nova,* 2002; Love 2000.
5. MacIntyre 1999:73; Van Tilburg and Ralston 2005.
6. We now have evidence in measures of center of mass that statues were reworked once they arrived at their *ahu* destinations, thus Van Tilburg's "statistically average *moai*" from Ahu Akivi reflects a statue already erected on the platform and not those in transit abandoned along the ancient roads. Center of mass is critical to transporting the statues, so ignoring these modifications by using a reshaped *moai* that had already been erected on an *ahu* platform raises doubts about conclusions that follow from such experiments or simulations.

7. Lee 1999.
8. Routledge 1919:195.
9. Counts of *moai* found faceup or facedown relative to the slope along the road heading away from the Rano Raraku quarry.

Direction of Slope Relative to *Moai*

	Downhill	**Flat**	**Uphill**	**Total**
Facedown	17	3	11	31
Faceup	3	0	16	19
Moai Position				
Lateral	0	0	1	1
Total	20	3	28	51

Chi-square = 12.388, p = 0.01476

10. Modeling *moai* form took advantage of Photosynth software, produced by Microsoft, which allows one to generate a three-dimensional representation directly from photographs.

CHAPTER SIX.
A PEACEABLE ISLAND

1. Owsley et al. 1994.
2. Routledge 1919:223.
3. In fact, obsidian blades are so sharp that some doctors consider them superior tools for eye surgery. Sheets 1989.
4. Thomas and Berghof 2000:304.
5. Bahn and Flenley 1992:165; Diamond 2005:109; Métraux 1940:166–167, 376; Thomson 1889:532.
6. Not a bad starting assumption given that *mata'a* are the only candidate we have for systematically constructed weaponlike objects.
7. Church and Rigney 1994; Church and Ellis 1996.
8. As translated and quoted from Bouman's journal by Von Saher 1994:100.
9. In 1891, William J. Thomson noted that weapons among the islanders included "throwing stones," though he also noted that slings were not used.
10. Ruiz-Tagle 2005:63.
11. Ferdon 1961:9–22.
12. Reanier and Ryan 2003.
13. Sir Conan Doyle, *The Valley of Fear* (1915).
14. Froude 1886.
15. Twain 1906.
16. Maynard Smith and Price 1973.
17. Corney et al. 1908.

NOTES

CHAPTER SEVEN.
AHU AND HOUSES

1. The use of the term *evolution* in *cultural evolution* often confuses many who think this is the same sort of evolution that explains biological change—Darwinian evolution. Nothing could be further from the truth. The old school of cultural evolution assumes that change is orthogenetic: that change increases in a progressive fashion toward "higher levels" of evolution. This kind of pattern is taken to be self-evident: in the past, people and organisms were "simpler," and now they are more complex. Of course, this scaling of observation depends on describing the world in scaled terms (that is, using terms that allow you to array observations in an order). This way of describing the world is not necessary, nor meaningful. In Darwinian evolution, change is simply the result of variation produced and sorted: life changes as a consequence of variability and contingent constraints.
2. Moore 1990.
3. Brown 1924.
4. Martinsson-Wallin 1994.
5. Cook 1846:253.
6. Small chunks of coral are often found on *ahu* and may be related to ceremonial activity since they must have been collected from near-shore environments and do not naturally occur in island deposits. Corals grow in the shallow waters near Rapa Nui, though the diversity of coral species is quite low relative to other parts of the Pacific. Only thirteen species of coral are known to live near the island, six of which are in the genus *Pocillopora* (Glynn et al. 2007).
7. Ruiz-Tagle 2005:31.
8. Ruiz-Tagle 2005:59.
9. Ruiz-Tagle 2005:59.
10. La Pérouse in Métraux 1940:344.
11. Routledge 1919:233.
12. Heyerdahl and Skjølsvold 1961.
13. Beechey 1831.
14. Routledge 1919.
15. Lavachery 1936.
16. Whittle 1987.
17. McCoy 1976.
18. Ruiz-Tagle 2005:34.
19. Anthropologists often describe the archaeological record in terms of features meant for the "elite" versus those used by "commoners." The use of *elite* here suggests some sort of differential access to power or status for individuals. Deconstructing this notion, however, shows that it is an empty and baseless claim. The idea used to separate "elite" from "commoner" holds that fancy things (such as elaborate house constructions, exotic goods) symbolize "elite" status. Unfortunately, such an assertion is circular since fancy things point to elite status and elite status is demon-

strated by the presence of fancy things. Fundamentally, the identification is simply one related to the composition and/or investment made in the artifact and nothing more.

20. Lipo et al. 2010.
21. Gill 1988; Gill et al. 1983.
22. Dudgeon 2008.
23. Stefan 1999.

CHAPTER EIGHT. THE BENEFITS OF MAKING *MOAI*

1. Of course, these "heads" are full-sized statues with bodies, but the buried and standing statues of Rano Raraku continue to lead people (students and otherwise) to describe the *moai* as "heads."
2. Veblen 1899/1994.
3. Smith and Bliege Bird 2000; Bliege Bird and Smith 2005.
4. Moore 1957.
5. Moore 1957:72.
6. During the Spanish visit to Rapa Nui in 1770, Don Francisco Antonio de Agüera y Infanzon, chief pilot of the Spanish frigate *Santa Rosalia,* noted the lower number of women to men. On the same expedition, First Pilot Don Juan Hervé of the *San Lorenzo* also remarked that "the number of inhabitants, including both sexes will be from about nine hundred to a thousand souls: and of these very few are women—I do believe they amount to seventy—and but few boys" (Ruiz-Tagle 2005:93). In 1774, Cook also noted this issue and wrote, "They either have but few females among them, or many were restrained from making their appearance, during our stay" (Ruiz-Tagle 2005:167). In 1786, La Pérouse observed that among 1,200 individuals who gathered on the beach, not more than three hundred were female (La Pérouse 1968). Like Cook, La Pérouse suspected that women were elsewhere (that is, in the countryside or caves).
7. Stefan 2000:69. We need to make one caveat about skeletal remains. Like many aspects of the archaeological record, often there is ambiguity between what we find and what we would like to know. In the case of skeletal remains, differences in cultural treatment between males and females might also explain the differences in sex. It is possible that men were more likely to be buried in ways that led to their discovery. Or by chance alone, our collections may have the sex ratio distribution they do. In the case of Rapa Nui, there does not appear to be evidence of differential treatment of men and women in burials, at least as far as burial goods and locations are concerned. In addition, the sample is relatively large, so these kinds of biases seem less likely. Finally, the corroboration of the sex ratio with historic accounts points to this difference being a reliable inference about the prehistoric population as well.
8. Polet 2006.
9. Williams 1966.

10. Sober and Wilson 1998.
11. Maynard Smith 1964.

CHAPTER NINE. THE COLLAPSE

1. Corney et al. 1908:8.
2. Corney et al. 1908:11.
3. Corney et al. 1908:12.
4. Ruiz-Tagle 2005:158.
5. Ruiz-Tagle 2005:157.
6. La Pérouse 1968:320.
7. La Pérouse 1968:325.
8. British archaeologists Colin Richards and Sue Hamilton from the University of Manchester and University College London made headlines in 2009 for their research on the *pukao* quarried at Puna Pau. The future planned work by this research team to excavate areas around Puna Pau is likely to reveal details about the use of the quarry throughout prehistory.
9. Ruiz-Tagle 2005:162.
10. Corney et al. 1908:94.
11. Richards 2008.
12. Routledge 1919:239.
13. Routledge 1919:239.
14. La Pérouse 1968:328.
15. Corney 1908:25; also described in Fischer 2005:52.
16. Fischer 2005:52.
17. Diamond 1997:196–197.
18. Wolfe et al. 2007.
19. Ewald 1994.
20. Dobyns 1983, 1993; Stannard 1989.
21. In 1854, English physician John Snow traced the source of cholera infections in London to a communal pump that was used for obtaining drinking water by infected individuals. Locking the pump, Snow was able to stop a major disease outbreak. His finding demonstrated that cholera was not spread through "miasma" or "bad air" but by contaminated water.
22. Ruiz-Tagle 2005:54.
23. Ruiz-Tagle 2005:57.
24. Ruiz-Tagle 2005:63, 61.
25. Thomas and Berghof 2000:300.
26. Ruiz-Tagle 2005:155–156.
27. Ruiz-Tagle 2005:155–156.
28. Thomas and Berghof 2000:300, 302.
29. Ruiz-Tagle 2005:156.
30. Ruiz-Tagle 2005:157.
31. Thomas and Berghof 2000:303–304.
32. Thomas and Berghof 2000:315.

33. Thomas and Berghof 2000:318.
34. This account of dining in the shade of the statue suggests it may have been standing at the time. The area they are describing is almost certainly the interior of the southern coast along the statue roads that emanate from the quarry at Rano Raraku; Thomas and Berghof 2000:319.
35. Thomas and Berghof 2000:324.
36. Thomas and Berghof 2000:308.
37. La Pérouse 1968:327.
38. Métraux 1940.
39. Fischer 2005:75.
40. Fischer 2005:75.
41. Fischer 1992:73.
42. Bahn 1997:123.
43. Fischer 2005:79.
44. Fischer 2005:80.
45. Fischer 2005:88.
46. Fischer 2005:90.
47. Fischer 2005:91.
48. Fischer 2005:188.
49. Fischer 2005:91–92.
50. Fischer 1997.
51. Altman 2004:50.
52. Fischer 2005:110–119.
53. Thomson 1889:461.
54. Ayres and Ayres 1995:49.
55. Thomson 1889:458.
56. Thomson 1889:456.
57. Thomson 1889:454.
58. Thomson 1889:459.
59. Routledge 1919:141–142.
60. Fischer 2005:168.
61. Routledge 1919:212.
62. http://berkeley.edu/news/media/releases/2002/11/peace/rapu.html.

CHAPTER TEN.
CONCLUSION

1. The name Rapa Iti is used to distinguish this island from Rapa Nui. Rapa Iti means "Little Rapa" while Rapa Nui means "Big Rapa." This distinction is accurate, as Rapa Iti is only 15 square miles while Rapa Nui is 63 square miles.
2. Kennett et al. 2006; Ferdon 1961.
3. The inhabitants of Rapa Nui are well aware of their situation and are struggling to find a way of handling it. Beginning in August of 2009, Rapa Nui protesters began to periodically occupy the island's airport runway and other government facilities in an attempt to bring attention to the island's

difficulties. The problem is obviously not a simple one since the potential solutions involve trade-offs. On the one hand, success on the island depends on tourism and the inflow of resources from other parts of the world. As a result of increased tourism, the island's native residents have achieved relative prosperity through businesses and land rentals to Chilean enterprises. Any limits on growth would directly restrain the success of individuals.

On the other hand, growth has consequences for such shared resources as water, electricity, and food supply. Growth also brings direct impacts to the archaeological record. An increasing number of tourists has resulted in increased erosion in fragile areas like Orongo, where many of the famous birdman petroglyphs cling precariously about precipitous sea cliffs. At the quarry area, only marked trails can be traversed now, as previous explorations have caused erosion to the steep hill slopes. More direct damage has also occurred. In the summer of 2008, a Finnish tourist broke an earlobe from a statue on the island, causing much public outcry.

The destruction of the archaeological record is not just limited to the actions of visitors. Motivated by new sources of income, many islanders are choosing to lease land that they own to Chilean firms. A recently constructed Chilean-owned, five-star hotel, for example, is located on land that once was used as pasture. This thousand-dollar-a-night hotel is situated on a landscape that was once, like the rest of the island, littered with archaeological remains. In addition, much of the interior parts of the island are being returned from the Chilean government to individual native ownership. While this transfer of land recognizes the centuries of abuse suffered by native Rapanui and the unlawful seizure of the land from its inhabitants, each time a parcel of land is returned to private hands, the archaeological record is potentially threatened. Many landowners establish houses or bulldoze the rocky surface to create small fields for crops. While each owner may appreciate the overall value of the archaeological record as local heritage and source of tourism, personal incentives all too often result in loss of the archaeological record.

The current situation is an excellent example of the "tragedy of the commons." As a group, everyone understands the fragile nature of the island, the limits of its infrastructure, and the value of the archaeological record. Yet, each individual tends to make decisions to best benefit him- or herself. Tourists seek exclusive access to less-seen parts of the island. Business owners want to bring in more tourists and desire more facilities to house and feed these visitors. Individuals want their portion of the tourist trade and land to enable them to follow their own ambitions. Everyone benefits from an increased number of flights, a bigger airport, and additional services. At the same time, such growth results in direct impact to the island's resources and the integrity of the archaeological record.

NOTES

APPENDIX 1.
ENVIRONMENTAL CONSTRAINTS

1. Orographic effects occur when air in the atmosphere is forced upward due to topography such as hills or mountains. As the air gains altitude it expands and cools. This process can increase relative humidity to 100 percent and create clouds and, often, precipitation.
2. Addison 2008.
3. McArthur 1982.
4. Some researchers on Rapa Nui have claimed that rainfall is higher under El Niño–Southern Oscillation (ENSO) conditions; others have argued it brings drought. To address the conflicting speculations, Genz and Hunt (2003) analyzed ENSO years ranked by their strength against the available rainfall data and found that no correlation can be detected. Rapa Nui rainfall, based on modern data, is variable but independent of ENSO. Climate and rainfall on Rapa Nui likely vary with other regional climatic trends, such as the Interdecadal Pacific Oscillation (a roughly 20–30 year cycle of warming and cooling).
5. Based on FAO data.
6. These are modern statistics, the only detailed information available, and they paint one picture of agriculture on prehistoric Rapa Nui. We know that global climate has varied both in time and geographically. Archaeologist and bestselling author Brian Fagan (2001, 2009) has detailed the effects of the Medieval Warm Period and the Little Ice Age, but the effects beyond increased ENSO frequencies in the South Pacific remain open to further investigation. Melinda Allen (2006) has discussed the possibility that the Little Ice Age, a period of cooler temperatures in many parts of the world, may have had the opposite effects in the South Pacific, with conditions that were warmer and wetter, yet stormier, including perhaps on Rapa Nui. Warmer and wetter conditions likely made agricultural pursuits more productive, but increased storminess would have brought its own problems, including high winds.
7. Louwagie et al. 2006.
8. Louwagie et al. 2006.
9. Diamond 2005.

APPENDIX 2.
LITHIC MULCHING AND *MANAVAI*

1. Lightfoot 1993a, 1993b, 1994, 1997; Lightfoot and Eddy 1995.
2. Lightfoot 1997.
3. Stevenson et al. 2009.
4. Stevenson and Wozniak 1999; Stevenson et al. 2002.
5. Louwagie and Langohr 2003; Louwagie et al. 2006.
6. Wofford 2006.

Bibliography

Addison, D.J. 2008. The changing role of irrigated Colocasia esculenta (taro) on Nuku Hiva, Marquesas Islands: From an essential element of colonization to an important risk-reduction strategy. *Asian Perspectives* 47:139–155.

Allan, J.A., Richards, T.S. 1983. Use of satellite imagery in archaeological surveys. *Annual Report for Libyan Studies* 14:4–8.

Allen, M.S. 2006. New ideas about Late Holocene climate variability in the Central Pacific. *Current Anthropology* 47:521–535.

Altman, A.M. 2004. *Early Visitors to Easter Island 1864–1877.* Easter Island Foundation, Los Osos, Calif.

Anderson, A. 1991. The chronology of colonization in New Zealand. *Antiquity* 65:767–795.

Anderson, A. 1995. Current approaches in East Polynesia colonisation research. *Journal of the Polynesian Society* 104:110–132.

Anderson, A. 1996. Was *Rattus exulans* in New Zealand 2000 years ago? AMS radiocarbon ages from Shag River Mouth. *Archaeology of Oceania* 31:178–184.

Anderson, A. 2003. Entering uncharted waters: Models of initial colonization in Polynesia. In Rockman, M., Steele, J. (Eds.), *Colonization of unfamiliar landscapes: The Archaeology of Adaptation.* Routledge, London, pp. 169–205.

Anderson, A. 2009. The rat and the octopus: Initial human colonization and the prehistoric introduction of domestic animals to Remote Oceania. *Biological Invasions* 11:1503–1519.

Anderson, A. 2009. Epilogue: Changing archaeological perspectives upon historical ecology in the Pacific Islands. *Pacific Science* 63:747–757.

Anovitz, L., Elam, J.M., Riciputi, L., Cole, D. 1999. The failure of obsidian hydration dating: Sources, implications, and new directions. *Journal of Archaeological Science* 26:18.

Athens, J.S. 2009. *Rattus exulans* and the catastrophic disappearance of Hawai'i's native lowland forest. *Biological Invasions* 11:1489–1501.

Athens, J.S., Tuggle, H.D., Ward, J.V., Welch, D.J. 2002. Avifaunal extinctions, vegetation change, and Polynesian impacts in prehistoric Hawai'i. *Archaeology in Oceania* 37:57–78.

BIBLIOGRAPHY

Auld, T.D., Hutton, I., Ooi, M.K.J., Denham, A.J. 2010. Disruption of recruitment in two endemic palms on Lord Howe Island by invasive rats. *Biological Invasions* 12:3351–3361.

Ayres, W.S., Ayres, G.S. (Translators) 1995. *Geiseler's Easter Island report: An 1880s anthropological account.* Asian and Pacific Archaeology Series No. 12. Social Science Research Institute, University of Hawaii, Honolulu.

Bahn, P. 1997. Easter Island or (Man-) Eaters Island? *Rapa Nui Journal* 11:123–125.

Bahn, P., Flenley, J. 1992. *Easter Island, Earth Island.* Thames & Hudson, London.

Barclay, H.V. 1899. Easter Island and its colossal statues. *Royal Geographical Society of Australasia, South Australia Branch, Proceedings* 3:127–137.

Barnes, S., Matisoo-Smith, L., Hunt, T.L. 2006. Ancient DNA of the Pacific rat (*Rattus exulans*) from Rapa Nui (Easter Island). *Journal of Archaeological Science* 33:1111.

Barthel, T.S. 1958. Female stone figures on Easter Island. *Journal of the Polynesian Society* 67:252–255.

Barthel, T.S. 1978. *The eighth land: Polynesian discovery and settlement of Easter Island.* Honolulu: University of Hawaii Press.

Beechey, F.W. 1831. *Narrative of a voyage to the Pacific and Bering's Strait.* Carey & Lea, Philadelphia.

Behrens, C.F. 1903. Another narrative of Jacob Roggeveen's visit. In Corney, B.G. (Ed.), *The voyage of Captain Don Felipe González in the ship of the line San Lorenzo, with the frigate Santa Rosalia in company, to Easter Island in 1770–1: Preceded by an extract from Mynheer Jacob Roggeveen's official log of his discovery of and visit to Easter Island, in 1722. Kraus Reprint, Nendeln, Liechtenstein.*

Birdsell, J.B. 1957. Some population problems involving Pleistocene man. *Cold Spring Harbor Symposia on Quantitative Biology* 22:47–69.

Bliege Bird, R., Smith, E.A., 2005. Signaling theory, strategic interaction and symbolic capital. *Current Anthropology* 6:221–248.

Boone, J.L. 1998. The evolution of magnanimity: When is it better to give than to receive? *Human Nature* 9:1–21.

Boone, J.L. 2000. Social power, status, and lineage survival. In Diehl, M.W. (Ed.), *Hierarchies in action: Cui bono?* Southern Illinois University Press, Carbondale.

Boone, J.L., Kessler, K.L. 1999. More status or more children? Social status, fertility reduction, and long-term fitness. *Evolution and Human Behavior* 20:257–277.

Bork, H.-R., Mieth, A. 2003. The key role of *Jubaea* palm trees in the history of Rapa Nui: A provocative interpretation. *Rapa Nui Journal* 17:119–121.

Bork, H.-R., Mieth, A., Tschochner, B., 2004. Nothing but stones? A review of the extent and technical efforts of prehistoric stone mulching on Rapa Nui. *Rapa Nui Journal* 18:10–14.

Bridgman, H.A. 1983. Could climate change have an influence on the Polynesian migrations? *Palaeogeography, Palaeoclimatology, Palaeoecology* 41:193–206.

Brown, J.M. 1924. *The riddle of the Pacific.* T. Fisher Unwin, London.

Buck, P.H. 1938. *Vikings of the sunrise.* J.B. Lippincott, Philadelphia.

Burley, D.V., Dickinson, W.R. 2004. Late Laptia occupation and its ceramic assemblage at the Sigatoka Sand Dune site, Fiji, and their place in Oceanic prehistory. *Archaeology in Oceania* 39:12–25.

Burney, D.A. 2010. *Back to the future in the caves of Kaua'i: A scientist's adventures in the dark.* Yale University Press, New Haven.

Burney, L.P., Burney, D.A. 2003. Charcoal stratigraphies for Kaua'i and the timing of human arrival. *Pacific Science* 57:211–226.

Butler, K., Flenley, J.R. 2001. Further pollen evidence from Easter Island. In Stevenson, C.M., Lee, G., Morin, F.J. (Eds.), *Pacific 2000: Proceedings of the Fifth International Conference on Easter Island and the Pacific,* pp. 79–86.

Butler, K., Flenley, J.R. 2010. The Rano Kau 2 pollen diagram: palaeoecology revealed. *Rapa Nui Journal* 24:5–10.

Butler, K., Prior, C.A., Flenley, J.R. 2004. Anomalous radiocarbon dates from Easter Island. *Radiocarbon* 46:395–405.

Caborn, J.M. 1957. *Shelterbelts and microclimate.* Forestry Commission Bulletin 209. Forestry Commission, Edinburgh.

Campbell, D.J., Atkinson, I.A.E. 2002. Depression of tree recruitment by the Pacific rat (*Rattus exulans* Peale) on New Zealand's northern offshore islands. *Biological Conservation* 107:19–35.

Chapman, P.M. 1996. The role of biological anthropology in Easter Island research. *Rapa Nui Journal* 10(3):53–56.

Chapman, P.M. 1997. A biological review of the prehistoric Rapanui. *Journal of the Polynesian Society* 106:161–174.

Charnov, E.L. 1982. *Sex allocation.* Princeton University Press, Princeton, N.J.

Church, F., Ellis, G. 1996. A use-wear analysis of obsidian tools from an Ana Kionga. *Rapa Nui Journal* 10:81–88.

Church, F., Rigney, J. 1994. A microwear analysis of tools from site 10-241, Easter Island—An inland processing site. *Rapa Nui Journal* 8:101–105.

Cook, J. 1777. *A voyage towards the South Pole and round the world, 1772–75.* London.

Cook, J. 1846. *The voyages of Captain James Cook round the world.* William Smith, London.

Corney, B.G., González de Haedo, F., Roggeveen, J. 1908. *The voyage of Captain Don Felipe González in the ship of the line San Lorenzo, with the frigate Santa Rosalia in company, to Easter Island in 1770–1. Preceded by an extract from Mynheer Jacob Roggeveen's official log of his discovery of and visit to Easter Island, in 1722.* Hakuyt Society, Cambridge.

Custer, J.F. 1986. Application of LANDSAT data and synoptic remote sensing to predictive models for prehistoric archaeological sites: An example from the Delaware coastal plain. *American Antiquity* 53:572–588.

Dancey, W.S., Pacheco, P.J. 1997. *Ohio Hopewell Community Organization.* Kent State University Press, Kent, Ohio.

Davidson, J. 1984. *The prehistory of New Zealand.* Longman Paul, Auckland.

Diamond, J. 1985. Rats as agents of extermination. *Nature* 318:602–603.

Diamond, J. 1995. Easter Island's end. *Discover* 9:62–69.

Diamond, J. 1997. *Guns, germs, and steel: The fates of human societies.* Norton, New York.

Diamond, J. 2005. *Collapse: How societies choose to fail or succeed.* Viking, New York.

Diamond, J. 2007. Easter Island revisited. *Science* 317:1692–1694.

DiSalvo, L.H., Randall, J.E. 1993. The marine fauna of Rapanui, past and present. In Fischer, S.R. (Ed.), *Easter Island studies: Contributions to the history of Rapa Nui in memory of William T. Mulloy.* Oxbow Monograph.

DiSalvo, L.H., Randall, J.E., Cea, A. 1988. Ecological reconnaissance of Easter Island sublittoral marine environment. *National Geographic Research* 4:451–473.

Dobyns, H.F. 1983. *Their number become thinned.* University of Tennessee Press, Knoxville.

Dobyns, H.F. 1993. Disease transfer at contact. *Annual Review of Anthropology.* 22:273–291.

Doyle, A.C. 1915. *The valley of fear.* A.L. Burt, New York.

Drake, D.R., Hunt, T.L. 2009. Invasive rodents on islands: Integrating historical and contemporary ecology. *Biological Invasions* 11:1483–1487.

Dransfield, J., Flenley, J.R., King, S.M., Harkness, D.D., Rapu, S. 1984. A recently extinct palm from Easter Island. *Nature* 312:750–752.

Dudgeon, J.V. 2008. The genetic architecture of the Late Prehistoric and Protohistoric Rapa Nui (Easter Islanders). Ph.D. diss., University of Hawaii, Manoa.

Dunnell, R.C. 1980. Evolutionary theory and archaeology. In M.B. Schiffer (Ed.), *Advances in archaeological method and theory.* Academic Press, New York, pp. 35–99.

Dunnell, R.C. 1982. Science, social science, and common sense: The agonizing dilemma of modern archaeology. *Journal of Anthropological Research* 38:1–25.

Dunnell, R.C. 1988. The concept of progress in cultural evolution. In Nitecki, M.H. (Ed.), *Evolutionary progress.* University of Chicago Press, Chicago, pp. 169–194.

Dunnell, R.C. 1989. Aspects of the application of evolutionary theory in archaeology. In Lamberg-Karlovsky, C.C. (Ed.), *Archaeological thought in America.* Cambridge University Press, Cambridge, pp. 35–49.

Dunnell, R.C. 1999. The concept of waste in an evolutionary archaeology. *Journal of Anthropological Archaeology* 18:243–250.

Dunnell, R.C., Greenlee, D. 1999. Late Woodland period "waste" reduction in the Ohio River Valley. *Journal of Anthropological Archaeology* 18:376–395.

Ebert, J.I., Lyons, T.R. 1980. Prehistoric irrigation canals identified from Skylab III and Landsat imagery in Phoenix, Arizona. In Lyons, T.R., Mathien, F.J. (Eds.), *Cultural resources remote sensing.* Cultural Resources Management Division, National Park Service, Washington, D.C., pp. 209–228.

Edwards, E., Marchetti, R., Dominichetti, L., Gonzales-Ferran, O. 1996. When the earth trembled, the statues fell. *Rapa Nui Journal* 10:1–15.

Emory, K.P. 1972. Easter Island's position in the prehistory of Polynesia. *Journal of the Polynesian Society* 81:57–89.

Emory, K.P., Sinoto, Y. 1965. *Preliminary report on the archaeological investigations in Polynesia: Field work in the Society and Tuamotu Islands, French Polynesia, and American Samoa in 1962, 1963, 1964.* Bishop Museum, Honolulu.

Englert, S. 1974. *La tierra de Hotu Matu'a: Historia, etnologia, y lengua de la Isla de Pascua.* Padre de Casas, Santiago.

Ewald, P.W. 1994. *Evolution of infectious disease.* Oxford University Press, Oxford.

Fagan, B. 2001. *The little Ice Age: How climate made history, 1300–1850.* Basic Books, New York.

Fagan, B. 2009. *The great warming: Climate change and the rise and fall of civilizations.* Bloomsbury Press, New York.

Failmezger, V. 2001. High resolution aerial color IR, multi-spectral, hyper-spectral, and SAR imagery over the Oatlands Plantation archaeological site near Leesburg—a pilot study of the new high resolution satellite imagery. In Campana, S., Forte, M. (Eds.), *Remote sensing in archaeology.* All'Insegna Del Giglio, Florence, pp. 143–148.

Fenchel, T. 1974. Intrinsic rate of natural increase: the relationship with body size. *Oecologica* 14:317–326.

Ferdon, E.N. 1961. Rapa Iti Island: A reconnaissance survey of three fortified hilltop villages. In Heyerdahl, T., Ferdon, E.N. (Eds.), *Norwegian archaeological expedition to Easter Island and the East Pacific, volume 2: Miscellaneous papers.* Rand McNally, New York, pp. 9–22.

Ferdon, E.N. 1981. A possible source of origin of the Easter Island boat-shaped house. *Asian Perspectives* 22:1–8.

Ferdon, E.N. 2000. Stone chicken coops on Easter Island. *Rapa Nui Journal* 14:77–79.

Finney, B. 1985. Anomalous westerlies, El Nino, and the colonization of Polynesia. *American Antiquity* 87:9–26.

Finney, B. 1993. Voyaging and isolation in Rapa Nui prehistory. *Rapa Nui Journal* 7:1–6.

Finney, B. 1994. The impact of Late Holocene climate change on Polynesia. *Rapa Nui Journal* 8:13–15.

Fischer, S.R. 1992. At the teeth of savages. *Rapa Nui Journal* 6:72–73.

Fischer, S.R. 1997. Rapanui's *Tu'u ko Iho* versus Mangareva's *'Atu Motua:* Evidence for multiple reanalysis and replacement in Rapanui settlement traditions, Easter Island. *Journal of Pacific History* 39:3–18.

Fischer, S.R. 2005. *Island at the end of the world: The turbulent history of Easter Island.* Reaktion Books, London.

Flenley, J.R. 1993. The palaeoecology of Rapa Nui, and its ecological disaster. In Fischer, S.R. (Ed.), *Easter Island studies: Contributions to the history of Rapanui in memory of William T. Mulloy.* Oxford Books, Oxford, pp. 27–45.

Flenley, J. 2010. A palynologist looks at the colonization of the Pacific. In Wallin, P., Martinsson-Wallin, H. (Eds.), *The Gotland papers: Selected papers from the VII International Conference on Easter Island and the Pacific.* Gotland University Press 11, Gotland, pp. 15–34.

Flenley, J., Bahn, P. 2002. *The enigmas of Easter Island.* 2nd ed. Oxford University Press, New York.

Flenley, J.R., King, S.M., 1984. Late Quaternary pollen records from Easter Island. *Nature* 307:47–50.

Flenley, J.R., King, S.M., Teller, J.T., Prentice, M.E., Jackson, J., Chew, C. 1991.

The Late Quaternary vegetational and climatic history of Easter Island. *Journal of Quaternary Science* 6:85–115.

Fowler, M.J.F. 1996. High-resolution satellite imagery in archaeological application: A Russian satellite photograph of the Stonehenge region. *Antiquity* 70:667–671.

Fowler, M.J.F. 2002. Satellite remote sensing and archaeology: A comparative study of satellite imagery of the environs of Figsbury Ring, Wiltshire. *Archaeological Prospection* 9:55–69.

Froude, J.A. 1886. *Oceana, or England and her colonies.* Charles Scribner, New York.

Genz, J., Hunt, T.L. 2003. El Niño/southern oscillation and Rapa Nui prehistory. *Rapa Nui Journal* 17:7–11.

Gibbons, A. 2001. The peopling of the Pacific. *Science* 291:1735–1737.

Gill, G. 1988. William Mulloy and the beginnings of Wyoming osteological research on Easter Island. *Rapa Nui Journal* 2:9–13.

Gill, G., Osley, D.W., Baker, S.J. 1983. Craniometric evaluation of prehistoric Easter Island populations. *American Journal of Physical Anthropology* 110:407–419.

Gintis, H. 2009. *The bounds of reason: Game theory and the unification of the behavioral sciences.* Princeton University Press, Princeton, N.J.

Giovas, C.M. 2006. No Pig Atoll: Island biogeography and the extirpation of a Polynesian domesticate. *Asian Perspectives* 45:69–95.

Glynn, P.W., Wellington, G.M., Riegl, B., Olson, D.B., Borneman, E., Wieters, E.A. 2007. Diversity and biogeography of the scleractinian coral fauna of Easter Island (Rapa Nui). *Pacific Science* 61:67–90.

Golson, J. 1965. Thor Heyerdahl and the prehistory of Easter Island. *Oceania* 36:33–38.

Gongora, J., Rawlence, N.J., Mobegi, V.A., Jianlin, H., Alcalde, J.A., Matus, J.T., Hanotte, O., Moran, C., Austin, J.J., Ulm, S., Anderson, A.J., Larson, G., Cooper, A., 2008. Reply to Storey et al.: More DNA and dating studies needed for ancient El Arenal-1 chickens. *Proceedings of the National Academy of Sciences* 105:E100.

Gould, S.J. 1978. Sociobiology: The art of story telling. *New Scientist* 16:530–533.

Gould, S.J. 1980. Shades of Lamarck. In *The Panda's thumb.* Norton, New York, pp. 76–84.

Grant, B.R., Grant, P.R. 1989. *Evolutionary dynamics of a natural population: The large cactus finch of the Galapagos.* University of Chicago Press, Chicago.

Grau, J. 1996. *Jubaea,* the palm of Chile and Easter Island? *Rapa Nui Journal* 10:37–40.

Green, R.C. 2000. Origins for the Rapanui of Easter Island before European contact: Solutions from holistic anthropology to an issue no longer much of a mystery. *Rapa Nui Journal* 14:71–76.

Gurley R.E., Liller, W. 1997. Palm tress, *mana,* and the moving of the *moai. Rapa Nui Journal* 11:82–84.

Hagelberg, E., Quevedo, S., Turbon, D., Clegg, J.B. 1994. DNA from ancient Easter Islanders. *Nature* 369:25–26.

Hamilton, W.D. 1963. The evolution of altruistic behavior. *American Naturalist* 97:354–356.

Hamilton, W.D. 1964. The genetical evolution of social behavior I. *Journal of Theoretical Biology* 7:1–16.

Hardin, G. 1968. The tragedy of the commons. *Science* 162:1243–1248.

Harris, M. 1959. The economy has no surplus? *American Anthropologist* 61:185–199.

Hather, J., Kirch, P.V. 1991. Prehistoric sweet potato (*Ipomoea batatas*) from Mangaia Island, Central Polynesia. *Antiquity* 65:887–893.

Hawkes, K. 1993. Why hunter-gatherers work: An ancient version of the problem of public goods. *Current Anthropology* 34:341–362.

Hawkes, K., Bliege Bird, R. 2002. Showing off, handicap signaling, and the evolution of men's work. *Evolutionary Anthropology* 11:58–67.

Heyerdahl, T. 1961. General discussion. In Heyerdahl, T., Ferdon, E. (Eds.), *Archaeology of Easter Island,* pp. 493–526.

Heyerdahl, T. 1976. *The art of Easter Island.* Allen & Unwin, London.

Heyerdahl, T. 1989. *Easter Island: The mystery solved.* Souvenir Press, London.

Heyerdahl, T., Ferdon, E. (Eds.) 1961. *Archaeology of Easter Island.* Allen & Unwin, London.

Heyerdahl, T., Skjølsvold, A. 1961. Notes on the archaeology of Pitcairn Island. In Heyerdahl, T., Ferdon, E. (Eds.) *Reports of the Norwegian archaeological expedition to Easter Island and the East Pacific,* vol. 2. Allen & Unwin, London, pp. 3–8.

Hill, K. 1993. Life history theory and evolutionary anthropology. *Evolutionary Anthropology* 2(3):78–88.

Holdaway, R.N. 1996. Arrival of rats in New Zealand. *Nature* 384(6606):225–226.

Houghton, P. 1996. *People of the great ocean: Aspects of human biology of the early Pacific.* Cambridge University Press, New York.

Hunt, T.L. 2006. Rethinking the fall of Easter Island: New evidence points to an alternative explanation for a civilization's collapse. *American Scientist* 94:412–419.

Hunt, T.L. 2007. Rethinking Easter Island's ecological catastrophe. *Journal of Archaeological Science* 34:485–502.

Hunt, T.L., Lipo, C.P. 2006. Late colonization of Easter Island. *Science* 311:1603–1606.

Hunt, T.L., Lipo, C.P. 2007. Chronology, deforestation, and "collapse": Evidence vs. faith in Rapa Nui prehistory. *Rapa Nui Journal* 21:85–97.

Hunt, T.L., Lipo, C.P. 2008. Evidence for a shorter chronology on Rapa Nui (Easter Island). *Journal of Island and Coastal Archaeology* 3:140–148.

Hunt, T.L., Lipo, C.P. 2008. Top-down archaeology: High resolution satellite images of Rapa Nui on Google Earth. *Rapa Nui Journal* 22:5–13.

Hunt, T.L., Lipo, C.P. 2009. Revisiting Rapa Nui (Easter Island) "Ecocide." *Pacific Science* 63:601–616.

Hunt, T.L., Lipo, C.P. 2009. Ecological catastrophe, collapse, and the myth of "ecocide" on Rapa Nui (Easter Island). In McAnany, P.A., Yoffee N. (Eds.), *Questioning collapse: Human resilience, ecological vulnerability, and the aftermath of empire.* Cambridge University Press, Cambridge, pp. 21–44.

Irwin, G. 1992. *The prehistoric exploration and colonization of the Pacific.* Cambridge University Press, Cambridge.

Jacob, F. 1998. *Of flies, mice, and men.* Harvard University Press, Cambridge, Mass.

Jakob, P.L. 1990. *Visions of a flying machine.* Smithsonian Institution Press, Washington, D.C.

Jones, D.E. 2007. *Poison arrows: North American Indian hunting and warfare.* University of Texas Press, Austin.

Kayser, M., Lao, O., Saar, K., Brauer, S., Wang, X., Nurnberg, P., Trent, R.J., Stoneking, M., 2008. Genome-wide analysis indicates more Asian than Melanesian ancestry of Polynesians. *American Journal of Human Genetics* 82:194–198.

Kennedy, D. 1998. Declassified satellite photographs and archaeology in the Middle East: Case studies from Turkey. *Antiquity* 72:553–561.

Kennett, D., Anderson, A., Prebble, M., Conte, E., Southon, J. 2006. Prehistoric human impacts on Rapa, French Polynesia. *Antiquity* 80:340–354.

Kirch, P.V. 1997. *The Lapita peoples.* Blackwell, Cambridge, England.

Kirch, P.V. 2000. *On the road of the winds: An archaeological history of the Pacific Islands before European contact.* University of California Press, Berkeley.

Kirch, P.V., Ellison, J. 1994. Palaeoenvironmental evidence for human colonization of remote Oceanic islands. *Antiquity* 68:310–321.

Kirch, P.V., Green, R.C. 2001. *Hawaiki, ancestral Polynesia: An essay in historical anthropology.* Cambridge University Press, Cambridge.

Kirch, P.V., Rallu, J.-L. (Eds.) 2007. *The growth and collapse of Pacific island societies: Archaeological and demographic perspectives.* University of Hawaii Press, Honolulu.

Klemmer, K., Zizka, G. 1993. The terrestrial fauna of Easter Island. In Fischer, S.R. (Ed.), *Easter Island studies: Contributions to the history of Rapa Nui in memory of William T. Mulloy.* Oxbow Monograph.

Krebs, J.R., Davies, N.B. 1981. *An introduction to behavioral ecology.* Alden Press, Oxford.

Ladefoged, T.N., Graves, M., McCoy, M. 2003. Archaeological evidence for agricultural development in Kohala, Island of Hawai'i. *Journal of Archaeological Science* 30:923–940.

Ladefoged, T.N., Stevenson, C.M., Vitousek, P., Chadwick, O. 2005. Soil nutrient depletion and the collapse of Rapa Nui society. *Rapa Nui Journal* 19:100–105.

La Pérouse, J.-F. 1968. *A voyage round the world performed in the years 1785, 1786, 1787, and 1788 by the* Boussole *and* Astrolabe. New York: Da Capo Press.

Lavachery, H. 1936. Contribution à l'étude de l'archéologie de l'ile de Pitcairn. *Bulletin de la Sociétés de Américanistes de Belgique* 13:3–42.

Lee, V.R. 1999. Rapa Nui rocks update. *Rapa Nui Journal* 13:16–17.

Lewontin, R. 1974. *The genetic basis of evolutionary change.* Columbia University Press, New York.

Lightfoot, D.R. 1993a. The cultural ecology of Puebloan pebble mulch. *Human Ecology* 21:115–143.

Lightfoot, D.R. 1993b. The landscape context of Anasazi pebble-mulched fields in the Galisteo Basin, northern New Mexico. *Geoarchaeology* 8:349–370.

Lightfoot, D.R. 1994. Morphology and ecology of lithic mulch agriculture. *Geographical Review* 84:172–185.

Lightfoot, D.R. 1997. The nature, history, and distribution of lithic mulch agriculture: An ancient technique of dryland agriculture. *Agricultural History Review* 44:206–222.

Lightfoot, D.R., Eddy, F.W. 1995. The construction and configuration of Anasazi pebble-mulch gardens in the northern Rio Grande. *American Antiquity* 60:12.

Lipo, C.P., Hunt, T.L. 2005. Mapping prehistoric statue roads on Easter Island. *Antiquity* 79:158–168.

Lipo, C.P., Hunt, T.L. 2009. AD 1680 and Rapa Nui prehistory. *Asian Perspectives* 48:309–317.

Lipo, C.P., Hunt, T.L., Hundtoft, B. 2010. Stylistic variability of stemmed obsidian tools (*mata'a*), frequency seriation, and the scale of social interaction on Rapa Nui (Easter Island). *Journal of Archaeological Science* 37(10):2551–2561.

Lloyd, W.F. 1833. *Two lectures on the checks to population.* Oxford.

Love, C.M. 2000. More on moving Easter Island statues, with comments on the *Nova* program. *Rapa Nui Journal* 14:115–118.

Love, C. M. 2001. The Easter Island *moai* roads. Western Wyoming Community College. Unpublished report on file at the P. Sebastian Englert Museum, Rapa Nui.

Louwagie, G., Langohr, R. 2003. Testing land evaluation methods for crop growth on two soils of the La Pérouse area (Easter Island, Chile). *Rapa Nui Journal* 17:23–27.

Louwagie, G., Stevenson, C., Langohr, R. 2006. The impact of moderate to marginal land suitability on prehistoric agricultural production and models of adaptive strategies for Easter Island (Rapa Nui, Chile). *Journal of Anthropological Archaeology* 25:290–317.

MacIntyre, F. 1999. Walking *moai? Rapa Nui Journal* 13(3):70–78.

MacIntyre, F. 2001. ENSO, Climate variability, and the Rapanui: II. Oceanography and Rapa Nui. *Rapa Nui Journal* 15:83–94.

Madsen, M., Lipo, C., Cannon, M. 1999. Fitness and reproductive trade-offs in uncertain environments: Explaining the evolution of cultural elaboration. *Journal of Anthropological Archaeology* 18:251–281.

Mann, D., Chase, J., Edwards, J., Beck, R., Reanier, R., Mass, M. 2003. Prehistoric destruction of the primeval soils and vegetation of Rapa Nui (Isla de Pascua, Easter Island). In Loret, J., Tanacredi, J.T. (Eds.), *Easter Island: Scientific exploration into the world's environmental problems in microcosm.* Kluwer Academic/Plenum, New York, pp. 133–153.

Mann, D., Edwards, J., Chase, J., Beck, W., Reanier, R., Mass, M., Finney, B., Loret, J. 2008. Drought, vegetation change, and human history on Rapa Nui (Isla de Pascua, Easter Island). *Quaternary Research* 69:16–28.

Martinsson-Wallin, H. 1994. *Ahu—the ceremonial stone structures of Easter Island: Analyses of variation and interpretation of meanings.* Societas Archaeologica Upsaliensis, Uppsala.

Martinsson-Wallin, H., Crockford, S.J. 2002. Early settlement of Rapa Nui (Easter Island). *Asian Perspectives* 40:244–278.

Maxwell, T.D. 1995. The use of comparative and engineering analyses in the study of prehistoric agriculture. In Teltser, P.A. (Ed.), *Evolutionary archaeology: Methodological issues.* University of Arizona Press, Tucson, pp. 113–128.

Maynard Smith, J. 1964. Group selection and kin selection. *Nature* 201:1145–1147.

Maynard Smith, J., Price, G.R. 1973. The logic of animal conflict. *Nature* 246 (2):15–18.

McAndrew, F.T. 2002. New evolutionary perspectives on altruism: Multilevel selection and costly signaling theories. *Current Directions in Psychological Science* 11:79–82.

McArthur, N. 1982. Isolated populations in enclaves or on small islands. In May, R.J., Nelson, H. (Eds.), *Melanesia: Beyond diversity.* Research School of Pacific Studies. The Australian National University, Canberra. pp. 27–32.

McCoy, P. 1976. *Easter Island settlement patterns in the Late Prehistoric and Protohistoric periods.* Easter Island Committee, International Fund for Monuments, New York.

McGlone, M.S., Wilmshurst, J.M. 1999a. A Holocene record of climate, vegetation change and peat bog development, East Otago, New Zealand. *Quaternary Science* 14:239–254.

McGlone, M.S., Wilmshurst, J.M. 1999b. Dating initial Maori environmental impact in New Zealand. *Quaternary International* 59:5–16.

Métraux, A. 1940. *Ethnology of Easter Island, Bulletin,* 160. Bernice P. Bishop Museum, Honolulu.

Métraux, A. 1957. *Easter Island: A Stone-Age civilization of the Pacific.* Oxford University Press, New York.

Mieth, A., Bork, H.-R. 2003. Diminution and degradation of environmental resources by prehistoric land use on Poike Peninsula, Easter Island (Rapa Nui). *Rapa Nui Journal* 17:34–42.

Mieth, A., Bork, H.-R. 2004. *Easter Island—Rapa Nui: Scientific pathways to secrets of the past.* Department of Ecotechnology and Ecosystem Development, Ecology Center, Christian-Albrechts-Universität zu Kiel, Kiel, Germany.

Mieth, A., Bork, H.-R. 2005. History, origin and extent of soil erosion on Easter Island (Rapa Nui). *Catena* 63:244–260.

Mieth, A., Bork, H.-R, Feeser, I. 2002. Prehistoric and recent land use effects on Poike Peninsula, Easter Island (Rapa Nui). *Rapa Nui Journal* 16:89–95.

Moore, O.K. 1957. Divination—A new perspective. *American Anthropologist* 59:69–74.

Moore, T. 1990. Thor Heyerdahl: Sailing against the current. *U.S. News & World Report,* April 2, 1990.

Mulloy, W. 1961. Ceremonial center of Vinapu. In Heyerdahl, T., Ferdon, E.N.J. (Eds.), *Archaeology of Easter Island.* Allen & Unwin, London, pp. 93–161.

Mulloy, W., Figueroa, G. 1978. *The A Kivi-Vai Teka complex and its relationship to Easter Island architectural prehistory.* Social Science Research Institute, University of Hawaii, Honolulu.

Mumford, G., Parcak, S. 2002. Satellite image analysis and archaeological fieldwork in El-Markha plain (south Sinai). *Antiquity* 76:353–354.

Nowak, M.A. 2006. *Evolutionary dynamics: Exploring the equations of life.* Belknap Press, Cambridge, Mass.

Nunn, P.D., Carson, R.H.-A.M.T., Thomas, F., Ulm, S., Rowland, M.J. 2007. Times of plenty, times of less: Last-millennium societal disruption in the Pacific basin. *Human Ecology.*

Omerod, P. 2005. *Why most things fail: Evolution, extinction, and economics.* Faber & Faber, London.

Oppenheimer, S. 2003. Austronesian spread into Southeast Asia and Oceania: Where from and when? In Sand, C. (Ed.), *Pacific archaeology: Assessments and prospects.* Service des Musées et du Patrimonie de Nouvelle-Calédonie, Nouméa, pp. 54–70.

Oppenheimer, S. 2004. The "express train" from Taiwan to Polynesia: On the congruence of proxy lines of evidence. *World Archaeology* 36:591–600.

Oppenheimer, S.J., Richards, M. 2001. Slow boat to Melanesia? *Nature* 410:166–167.

Orliac, C. 2000. The woody vegetation of Easter Island between the early 14th and the mid-17th centuries A.D. In Stevenson, C., Ayres, W. (Eds.), *Easter Island archaeology: Research on early Rapanui culture.* Easter Island Foundation, Los Osos, Calif., p. 211–220.

Orliac, C. 2003. Ligneux et palmiers de l'île de Pâques du Xiéme au XVIIéme siécle de notre ére. In Orliac, C. (Ed.), *Archéologie en Océanie insulaire: Peuplement, sociétés et paysages.* Editions Artcom, Paris, pp. 184–199.

Orliac, C., Orliac, M. 1998. The disappearance of Easter Island's forest: Overexploitation or climate catastrophe? In Stevenson, C.M., Ayers, W.S. (Eds.), *Easter Island in Pacific Context.* Easter Island Foundation, Los Osos, Calif., pp. 129–134.

Owsley, D.W., Gill, G.W., Ousley, S.D. 1994. Biological effects of European contact on Easter Island. In Larsen, C.S., Milner, G.R. (Eds.), *The wake of contact: Biological responses to conquest.* Wiley-Liss, New York, pp. 161–177.

Palmer, J.L. 1870. A visit to Easter Island, or Rapa Nui, in 1868. *Journal of the Royal Geographic Society* 40:167–181.

Pavel, P. 1995. Reconstruction of the transportation of the *moai* statues and *pukao* hats. *Rapa Nui Journal* 9:69–72.

Philip, G., Donoghue, D., Beck, A., Galiatsatos, N. 2002. CORONA satellite photography: An archaeological application from the Middle East. *Antiquity* 76:108–118.

Polet, C. 2006. Indicateurs de stress dans un échantillon d'anciens Pascuans. *Antropos* 11:261–270.

Pollan, M. 2001. *The botany of desire: A plant's eye view of the world.* Random House, New York.

Rainbird, P. 2002. A message for our future? The Rapa Nui (Easter Island) ecodisaster and Pacific Island environments. *World Archaeology* 33:436–451.

Reanier, R.E., Ryan, D.P. 2003. Mapping the Poike Ditch. In J. Loret and J.T. Tanacredi (Eds.), *Easter Island: Scientific exploration into the world's environmental problems in microcosm.* Kluwer Academic/Plenum Publishers, New York, pp. 133–221.

Reich, K.F. 1896. *Flora de Chile.* Cervantes, Santiago, Chile.

Richards, R. 2008. *Easter Island 1793 to 1861: Observations by early visitors before the slave raids.* Easter Island Foundation, Los Osos, Calif.

Richerson, P.J. 1977. Ecology and human ecology: A comparison of theories in the biological and social sciences. *American Ethnologist* 4:1–26.

Routledge, K. 1919. *The mystery of Easter Island.* Sifton, Praed, London.

Ruiz-Tagle, E. 2005. *Easter Island: The first three expeditions.* Museum Store, Rapanui Press, Rapa Nui.

Sahlins, M.M. 1955. Esoteric efflorescence in Easter Island. *American Anthropologist* 57:1045–1052.

Scaglion, R. 2005. Kumara in the Ecuadorian Gulf of Guayaquil? In Ballard, C., Brown, P. Bourke, R.M., Harwood, T. (Eds.), *The sweet potato in Oceania: A reappraisal.* University of Sydney Press, New South Wales, Australia, pp. 35–41.

Sellars, W. 1962. Philosophy and the scientific image of man. In Sellars, W. (Ed.), *Science, perception, and reality.* Routledge & Kegan Paul, London, pp. 1–40.

Sever, T.L., Wagner, D.W. 1991. Analysis of prehistoric roadways in Chaco Canyon using remotely sensed digital data. In Trombold, C. (Ed.), *Ancient road networks and settlement hierarchies in the New World.* Cambridge University Press, Cambridge, England, pp. 42–52.

Sheets, P. 1989. Dawn of a new Stone Age in eye surgery. In Poloefsky, A., Brown, P.J. (Eds.), *Applying anthropology: An introductory reader.* Mayfield, Mountain View, Calif., pp. 113–115.

Skjølsvold, A. 1994. *Archaeological investigations at Anakena, Easter Island.* Kon Tiki Museum Occasional Papers, Oslo.

Skjølsvold, A. 1996. Age of Easter Island settlement, Ahu and monolithic sculpture. *Rapa Nui Journal* 10:104–109.

Skottsberg, C. 1920. *The natural history of Juan Fernandez and Easter Island.* Almqvist & Wiksells Boktryckeri, Uppsala.

Smith, C.S. 1961a. A temporal sequence derived from certain ahu. In Heyerdahl, T., Ferdon, E.N. (Eds.), *Archaeology of Easter Island.* Allen & Unwin, London, pp. 181–218.

Smith, C.S. 1961b. The Poike Ditch. In Heyerdahl, T., Ferdon, E.N. (Eds.), *Archaeology of Easter Island.* Allen & Unwin, London, pp. 257–271.

Smith, E.A., Bliege Bird, R.L. 2000. Turtle hunting and tombstone opening: Public generosity as costly signaling. *Evolution and Human Behavior* 21:245–261.

Smith, E.A., Bowles, S., Gintis, H. 2000. Costly signaling and cooperation. *Journal of Theoretical Biology* 213:103–109.

Sober, E., Wilson, D.S. 1998. *Unto others: The evolution and psychology of unselfish behavior.* Harvard University Press, Cambridge, Mass.

Speck, F.G. 1935. *Naskapi: The savage hunters of the Labrador Peninsula.* University of Oklahoma Press, Norman.

Spriggs, M. 1989. The dating of the Island Southeast Asian Neolithic: An attempt at chronometric hygiene and linguistic correlation. *Antiquity* 63:587–613.

Spriggs, M., Anderson, A. 1993. Late colonisation of East Polynesia. *Antiquity* 67:200–217.

Stannard, D.E. 1989. *Before the horror: The population of Hawai'i on the eve of Western contact.* University of Hawaii Press, Honolulu.

Steadman, D. 2006. *Extinction and biogeography of tropical Pacific birds.* University of Chicago Press, Chicago.

Steadman, D.W., Casanova, P.V., Ferrando, C.C. 1994. Stratigraphy, chronology, and cultural context of an early faunal assemblage from Easter Island. *Asian Perspectives* 33:79–96.

Stefan, V.H. 1999. Craniometric variation and homogeneity in prehistoric/protohistoric Rapa Nui (Easter Island) regional populations. *American Journal of Physical Anthropology* 110:407–419.

Stefan, V.H. 2000. *Craniometric variation and biological affinity of the prehistoric Rapanui (Easter Islanders): Their origin, evolution, and place in Polynesian prehistory.* Ph.D. dissertation, University of New Mexico.

Stevenson, C.M., Abdelrehim, I.M., Novak, S.W. 2001. Infra-red photoacoustic and secondary ion mass spectrometry measurements of obsidian hydration rims. *Journal of Archaeological Science* 28:109–115.

Stevenson, C.M., Carpenter, J., Scheetz, B.E. 1989. Obsidian dating—Recent advances in the experimental determination and application of hydration rates. *Archaeometry* 31:193–206.

Stevenson, C.M., Haoa Cardinali, S., Ladefoged, T. 2009. *Easter Island (Rapa Nui) culture: Earthwatch 2009 expedition briefing.* http://www.earthwatch.org/Briefings/haoa_briefing.pdf, Earthwatch Institute, Maynard, Mass.

Stevenson, C.M., Ladefoged, T., Haoa, S. 2002. Productive strategies in an uncertain environment: Prehistoric agriculture on Easter Island. *Rapa Nui Journal* 16:17–22.

Stevenson, C.M., Wozniak, J. 1999. Prehistoric agricultural production on Easter Island (Rapa Nui), Chile. *Antiquity* 73:801.

Storey, A.A., Ramirez, J.M., Quiroz, D., Burley, D.V., Addison, D.J., Walter, R., Anderson, A.J., Hunt, T.L., Athens, J.S., Huynen, L., Matisoo-Smith, E.A. 2007. Radiocarbon and DNA evidence for a pre-Columbian introduction of Polynesian chickens to Chile. *Proceedings of the National Academy of Sciences* 104:10335–10339.

Suggs, R.C. 1961. *Archaeology of Nuku Hiva, Marquesas Islands, French Polynesia.* Anthropological Papers of the American Museum of Natural History 49, Part 1.

Terrell, J. 1988. *Prehistory in the Pacific Islands: A study of variation in language, customs, and human biology.* Cambridge University Press, Cambridge.

Terrell, J., Hunt, T.L., Gosden, C. 1997. The dimensions of social life in the Pacific: Human diversity and the myth of the primitive isolate. *Current Anthropology* 38:155–195.

Terrell, J.E., Kelly, K.M., Rainbird, P. 2001. Forgone conclusions? In search of Papuans and Austronesians. *Current Anthropology* 42:97–124.

Terrell, J.E., Welsch, R.L. 1997. Lapita and the temporal geography of prehistory. *Antiquity* 71:548–572.

Thomas, N., Berghof, O. (Eds.). 2000. *A voyage round the world: Georg Forster. Volume I.* University of Hawaii Press, Honolulu.

Thomson, W.J. 1889. Te Pito te Henua, or Easter Island, report of the U.S. National Museum for the year ending June 30, 1889. U.S. Government Printing Office, Washington, D.C., pp. 447–552.

Tomlinson, P.B. 2006. The uniqueness of palms. *Botanical Journal of the Linnean Society* 151:5–14.

Twain, M. 1906/1996. *What is man?* Oxford University Press, New York.

Ur, J. 2003. CORONA satellite photography and ancient road networks: A Northern Mesopotamia case study. *Antiquity* 77:102.

Van Tilburg, J.A. 1994. *Easter Island: Archaeology, ecology, and culture.* Smithsonian Institution Press, Washington, D.C.

Van Tilburg, J. 2003. *Among stone giants, the life of Katherine Routledge and her remarkable expedition to Easter Island.* Scribner, New York.

Van Tilburg, J., Ralston, T. 2005. Megaliths and mariners: experimental archaeology on Easter Island. In Johnson, K. (Ed.), *Onward and upward!: Papers in honor of Clement W. Meighan.* Stansbury Press, Chico, Calif., pp. 279–303.

Vargas Casanova, P. 1998. Rapa Nui settlement patterns: Types, function and spatial distribution of households structural components. In Vargas Casanova, P. (Ed.), *Easter Island and East Polynesian prehistory.* Instituto de Estudios Isla de Pascua, Facultad de Arquitectura y Urbanismo, Universidad de Chile, Santiago, pp. 111–130.

Vargas Casanova, P., Cristino, C., Izaurieta, R. 2006. *1000 años en Rapa Nui: Arqueología del asentamiento.* Instituto de Estudios Isla de Pascua, Universidad de Chile, Santiago, Chile.

Veblen, T. 1899/1994. *Theory of the leisure class.* Penguin Books, New York.

Vezzoli, L., Acocella, V. 2009. Easter Island, SE Pacific: An end-member type of hotspot volcanism. *Geological Society of America Bulletin* 121:869–886.

Von Saher, H., 1990/1991. Some details from the journal of Captain Bouman on the discovery of Easter Island. *Rapa Nui Journal* 4:49–52.

Von Saher, H., 1992. More journals on Easter Island: The works of Johann Reinhold Forster (1729–1798) and Johann Georg Adam Forster (1754–1794) [Part II]. *Rapa Nui Journal* 6:34–39.

Von Saher, H. 1994. The complete journal of Captain Cornelis Boumann from 31 March to 13 April 1722 during their stay around Easter Island. *Rapa Nui Journal* 8:95–100.

Whittle, A. 1987. Neolithic settlement patterns in temperate Europe: Progress and problems. *Journal of World Prehistory* 1:5–52.

Williams, G.C. 1966. *Adaptation and natural selection.* Princeton University Press, Princeton, N.J.

Wilmshurst, J.M., Anderson, A.J., Higham, T.F., Worthy, T.H. 2008. Dating the late prehistoric dispersal of Polynesians to New Zealand using the commensal Pacific rat. *Proceedings of the National Academy of Sciences* 105:7676–7680.

Wilmshurst, J.M., Hunt, T.L., Lipo, C.P., Anderson, A. 2011. High-precision radiocarbon dating shows recent and rapid initial human colonization of East Polynesia. *Proceedings of the National Academy of Science* 108:1815–1820.

Wilson, E.O. 1993. Is humanity suicidal? *Bio Systems* 31:235–242.

Wirtz, W.O. 1972. Population ecology of the Polynesian rat, *Rattus exulans,* on Kure Atoll, Hawaii. *Pacific Science* 26:433–464.

Wofford, G.B. 2006. Soil characteristics associated with agricultural enclosures (*manavai*) on Rapa Nui. Department of Global Earth Science, University of Hawaii, Honolulu.

Wolfe, N.D., Panosian Dunavan, C., Diamond, J. 2007. Origins of major human infectious diseases. *Nature* 447:279–28.

Wozniak, J.A. 1999. Prehistoric horticultural practices on Easter Island: Lithic mulched gardens and field systems. *Rapa Nui Journal* 13:95–99.

Yen, D.E. 1974. *The sweet potato and Oceania.* Bishop Museum Press, Honolulu.

Zahavi, A., Zahavi, A., 1997. *The handicap principle.* Oxford University Press, New York.

Acknowledgments

We thank the many individuals who contributed to the research that led to our writing this book, including the many University of Hawaii and California State University Long Beach archaeology field school students. We are also particularly grateful to the people of Rapa Nui who collaborated and contributed significantly to our efforts. The staffs of the P. Sebastian Englert Museum of Anthropology and the National Park (CONAF) have been tremendous hosts enabling our work. Although this list is far from complete, a number of individuals have been particularly helpful, and we thank them for their support and assistance: Atholl Anderson, J. Stephen Athens, Matthew Bell, Hannah Bloch, David Burney, Amy Commendador, Alan Cooper, John Dudgeon, the late Robert C. Dunnell, Enrique Pate Encina, Kelley Esh, Brian Fagan, Chris Filimoehala, Mark Ganter, Claudio Gomez, Francisco Torres Hochstetter, Tim Hunt, Bruce Kaiser, Marc Kelly, Jose Letelier, Emily Loose, Mark Madsen, Tsutomu Ben Matsumoto, Jacce Mikulanec, Claudia Peñafiel Morgan, Alex Morrison, Hector Neff, Susan Rabiner, Cecilia Rapu, Sergio Rapu, Timothy Reith, Deborah Schechter, John Terrell, Enrique Tucki, Geoffrey White, Janet Wilmshurst, and Gabriel Wofford.

Index

AD 1680 Event, 10–11
Addison, David, 182
Adena population, Illinois, 122, 135
Adventure (ship), 7
Agriculture. *See* Cultivation
Agüera y Infanzon, Francisco Antonio de, 96, 99, 106, 118, 119
Ahu. *See* Platforms (*ahu*)
Ahu Akahanga, 45, 56, 125
Ahu Akivi, 81
Ahu Ature Huke, 13
Ahu Hanga Roa, 113
Ahu Hanga Te'e o Vaihu, 56
Ahu Nau Nau, 13, 14, 116, 117, 127
Ahu Oroi, 56, 78
Ahu Tahai, 111
Ahu Tepeu, 114
Ahu Tongariki, 56, 111, 117, 143
Ahu Ura Uranga Te Mahina, 56
Ahu Vinapu, 113, 115
Akahanga, Easter Island, 58
Aku-Aku (Heyerdahl), 78
Alejandro Selkirk Island, 163
Altruism, 144–145
American Civil War, 165
Anakena Beach, Easter Island, 13–16, 34, 83, 117, 177
Anasazi, New Mexico, 43
Anderson, Atholl, 16
Anemia, 142
Angata (islander), 173
Animals
  cattle, 31, 60, 169, 170, 173
  dogs, 5, 34, 172
  horses, 31, 170, 171
  pigs, 5, 34, 170
  sheep, 31, 41, 46, 47, 52, 58, 59, 67, 169–174
Antarctic Circle, 7
Arend (ship), 147
Arizona, lithic mulching in, 43–44
Art of Easter Island, The (Heyerdahl), 10
Astrolabe (ship), 37, 42, 150
Athens, Steve, 25–29
Austral Islands, 4, 99, 120, 166, 167, 178
Australia, 7, 157

"Backrests," 3–4
Bahn, Paul, 20
Banana (*Musa sapientum*), 5, 21, 33, 35, 38, 40, 42, 160, 163, 183, 187, 188, 193, 194
Barclay, H. V., 147
Basalt stones, 112, 113, 124, 181, 192
Beechey, Frederick William, 120
Behrens, Carl Friedrich, 7, 156
Bet-hedging, benefits of, 134–142
Bird extinctions, 27, 28–29, 31
Blackbirding. *See* Slave raids
Bork, Hans-Rudolf, 46, 48
Bouman, Cornelius, 98, 155

Bounty (ship), 37, 120, 183, 184
Boussole (ship), 37, 150
Bradford, Ileana, 39–40
Breadfruit (*Artocarpus altilis*), 5, 182
Brown, John Macmillan, 33, 109, 110
Burial ceremonialism, 135
Burney, David, 190

Calcium, 196–198
Callao, Peru, 166
Canary Islands, 43
Cannibalism, 12, 165, 169
Canoes, 86, 161
Cargo cults, 153–155
Caribou hunting, 136–138
Carlos III, King of Spain, 158, 159
Catholicism, 166, 169
Cattle, 31, 60, 169, 170, 173
Caves, 99, 106, 166
Chadwick, Oliver, 47
Charcoal, 15–16, 23, 25, 26
Chickens
  on Easter Island, 5, 12, 21, 34, 171
  introduction of, 4
Chile, 19, 30–31, 163, 168, 170, 174, 175, 179
  rat impact in, 30–31
China
  lithic mulching in, 43
  scapulimancy in, 137
Chincha Islands, 166
Cholera, 156
Christmas Island, 38
Coconut (*Cocos nucifera*), 5, 182
Coconut palm, 21, 174
Collapse: How Societies Choose to Fail or Succeed (Diamond), 11–12
Colossus of Rhodes, 2
Conchoidal fracture, 85–86
Conspicuous consumption, 133
Cook, James, expedition to Easter Island (1774) by, 7, 33, 59, 78, 96–97, 113, 115, 141, 149, 151–153, 160–162
Cook Islands, 4, 120
Cooke, George, 171
Cooperative behavior, 144–146
Copeca, 119
Costly signaling, 132–134, 136, 143, 153
Cribra orbitalia, 142
Crocker, J., 163–164
Crop furrows, 51
Crops. *See* Plants
Cultivation, 33–53
  crop furrows, 51
  on Hawaiian Islands, 35, 36
  lack of, 35–37
  lithic mulching, 43–48, 50, 51, 191–197
  rock circles (*manavai*), 38–41, 50, 53, 123, 126, 140, 191–198
  rock gardens, 46
  shifting (slash-and-burn), 25, 27, 49–50, 53
  on slopes of Poike, 50–52
Cultural elaboration, 109, 110, 134–136, 140
Cultural evolution, 110
Cuming, Hugh, 42
Cupules, 124
Curbstones, 65–67, 77
Cuzco, Peru, 115

Darwin, Charles, 110, 144
Davis, Edward, 7, 158
Davis Island, 158
De Afrikaansche Galey (ship), 147, 148
Deforestation, 8, 11, 12, 17–32, 49, 189
Demon-Haunted World, The: Science as a Candle in the Dark (Sagan), 177
Dengue fever, 156
Diamond, Jared, 11–12, 20, 31, 156
DigitalGlobe, 67
Discoverer (ship), 42
Disease epidemics, 156–159, 161–164, 167, 169, 174
Dispersed-settlement pattern, 122–123, 126, 128, 135, 140, 178
Dogs, 5, 34, 172
Doyle, Sir Arthur Conan, 100

Dransfield, John, 30
Ducie Island, 94
Dudgeon, John, 128
Dupetit-Thouars, Abel Aubert, 153
Dutch expedition (1722), 2, 6, 10, 21, 24, 33–34, 36, 98, 118, 123–124, 147–149, 155, 158
Dutrou-Bornier, Jean-Baptiste, 169–170
Dysentery, 166

Early Period, 9, 10
Earth ovens (*umu*), 23, 123, 126, 127
Earthworks, 122, 135
Easter Island
  AD 1680 Event on, 10–11
  Anakena Beach, 13–16, 34, 117, 177
  ancient cultivation on, 38–42
  animals. *See* Animals
  birth of, 2
  cannibal label for, 165, 169
  caves on, 99, 106, 166
  Chile and, 163, 170, 174, 175, 179
  Christianization of, 168–169
  civil war on, 12, 50
  colonization of, 5–6, 15–17, 22, 24, 115, 143, 183
  competition between Long and Short Ears on, 10, 50
  conflict, lack of on, 94–101, 104–107
  Cook's expedition to (1774), 7–8, 33, 35, 59, 78, 96–97, 113, 115, 141, 149, 151–153, 160–163
  cultivation on. *See* Cultivation
  deforestation on, 8, 11, 12, 17–25, 29–32, 49, 189
  diet of islanders on, 34–35
  diseases brought to, 156–159, 161–164, 167, 169, 174
  Dutch expedition to (1722), 6, 10, 21, 24, 33–34, 36, 98, 118, 123–124, 147–149, 155, 158
  ecocide thesis on, 6, 11, 32, 146, 168
  first European sighting of (1722), 2, 6, 21, 147
  food shortages on, 142–143
  foods on. *See* Plants
  future of, 179–180
  Geiseler's visit to (1882), 59, 171
  Heyerdahl's research on, 9–11, 15–16
  houses on, 123–126
  insects on, 172
  isolation of, 3, 4
  Kon-Tiki Museum expedition to (1986), 79
  La Pérouse expedition to (1786), 8, 37, 39, 42–44, 119, 149–151, 155, 163
  landscape of, 2, 37, 56–57, 106, 181
  language and, 168
  marriage practices on, 127, 128
  missionaries on, 168–169
  nutritional deficiencies on, 142
  oral history of, 9–10, 13, 60
  periods of prehistory on, 9
  plants. *See* Plants
  population of, 6, 8, 11, 24, 32, 53, 60, 93–94, 105, 111, 141–143, 158, 167, 170–171, 174–175, 183
  rainfall on, 37, 47, 48, 181, 184–187, 189
  rats on, 29–31, 34, 35, 49, 51
  rebellion in, 60, 173
  roads on, 56–71, 77, 83, 84, 92
  rock circles (*manavai*) on, 38–41, 44, 50, 191–198
  Routledge and. *See* Routledge, Katherine
  Russian expedition to (1816), 164
  satellite images of, 39, 40, 51, 67–70
  seawall construction on, 112–115, 126
  sex ratio in population of, 141–142, 159, 161

Easter Island (*cont.*)
sexual encounters with visitors to, 159, 161, 163, 164, 169
size of, 5, 94, 105
skeletal remains on, 94, 127–128, 141, 142
slave raids and, 60, 106, 163–167
social organization on, 105–106, 111, 120–123, 145, 146
Spanish expedition to (1770), 6, 10, 34, 96, 99, 106, 118, 152, 158–159, 161, 162
statues on. *See* Statues (*moai*) of Easter Island
suicides on, 169
tourism and, 174, 175, 179
U.S. Air Force airstrip on, 174
volcanic origins of, 47–48, 51
water on, 3, 6, 162, 175, 181–182
winds and, 4–5, 187–189
Easter Island, Earth Island (Bahn and Flenley), 20
Easter Island palm (*Jubaea chilensis*), 11, 15, 19–24, 29–31, 34, 49, 51, 86
Ebola, 156
Ecocide thesis, 6, 11, 32, 146, 168
Edgecumbe, Lieutenant, 161, 162
Edmunds, Mr., 173
El Camino de los Moai, 57, 58
El Niño, 5
Ellen Snow (ship), 167
Enamel hypoplasia, 142
Endeavour (ship), 7
Englert, Father Sebastian, 11, 30
Escher, M. C., 75
Esh, Kelley, 13
Eucalyptus trees, 52, 81, 174
Evaporation, windbreaks and, 35
Evapotranspiration, 185–187
Evolutionary theory, 110, 132, 144
Ewa Plain, Oahu, 25–27
Eyraud, Eugène, 168

Fanning Island, 38
Farming. *See* Cultivation
Father Sebastian Englert Anthropological Museum, Easter Island, 96
Faulkner, William, 177
Female infanticide, 142
Figueroa, Gonzalo, 20
Fiji, 93, 106
Fire, in Hawaiian Islands, 26–27
Fischer, Steven Roger, 164–166, 167
Fish and seafood, 34, 179
Fleas, 172
Flenley, John, 20, 22, 30
Fleuriot de Langle, Paul Antoine, 42, 119
Flies, 172
Forster, Georg, 7, 35, 59, 160–163
Forster, Johann Reinhold, 7–8, 34, 161, 162
Fortifications, 99
Free rider problem, 144, 145
Froude, James A., 100
Fruits. *See* Plants

Gambier Islands, 4
Game theory, 101
Geiseler, Wilhelm, 59, 171
Genocide, 168
Germ theory, 156
Gill, George, 127
González de Haedo, Felipe, 118, 152, 158
Grave goods, 135
Gravitational potential energy, 89
Green, Roger, 44
Group selection, 144–146
Guano mines, 166

Hand axes (*toki*), 62, 73, 75, 77, 195
Hanga Roa, Easter Island, 41, 79, 160, 168–170, 173, 174
Haoa, Niko, 43
Haoa, Sonia, 44
Hard hammer percussion, 96
Hare-a-te-atua (ship-shaped house), 154
Hare paenga/hare vaka (boat-shaped house), 65, 124–125

Hats, importance of to islanders, 149–151, 153, 160
Hawaiian Islands, 2, 4, 17, 143, 165, 190
  deforestation of, 25–28
  farming and, 35, 36
  fighting among, 93
  lithic mulching in, 43
  stone statues on, 120
Hawks and Doves scenario, 101–105
Henderson Island, 4
Hervé, Juan, 118
Heyerdahl, Thor, 50, 61–62, 66, 75, 84, 174
  experiments with moving statues, 78, 86
  Pavel's experiment and, 79–80
  research on Easter Island, 9–11, 15–16
  theory on statues of, 110, 111, 115
Heyerdahl Expedition, 19–20
HIV (human immunodeficiency virus), 156
Hiva Oa Island, 120
Hodges, William, 7, 161
Hohokam, Arizona, 43
Hopewell population, Ohio, 122, 135
Horses, 31, 169, 170, 171
Hotu Matu'a, 5, 13
Houses, 123–126
Howea palm, 29
Howland Island, 38
Hyäne (ship), 59, 171

Inca, 82, 110, 111, 115
Indentures, 166, 174
Infanticide, 142
Influenza, 156
Insects, 172
Interbreeding, 145
International Archaeological Research Institute, Honolulu, 25
Inverse pendulum, physics of, 89
Irrigation, 182
Isla de San Carlos, 158
Israel, lithic mulching in, 43

Jaussen, Bishop, 166
Jubaea chilensis (large palm). *See* Easter Island palm.

Kava, 5, 182
Kentia palm, 29
Kin selection, 144
Kinetic energy of movement, 89–90
King, Sarah, 22
Kohala field system, Hawaiian Islands, 36
Kon-Tiki expedition, 9
Kon-Tiki Museum expedition (1986), 79
Kona Coast, Hawaiian Islands, 43
Korea, scapulimancy in, 137
Kotzebue, O. E., 164

La Campana National Park, Chile, 30
La Pérouse, Jean-François de Galaup, Comte de, expedition to Easter Island (1786) by, 8, 37, 39, 42–44, 119, 149–151, 155, 163
Labrador Peninsula, 136
Ladefoged, Thegn, 47
Langohr, Roger, 48, 193–194
Lanzhou, China, 43
Lapita pottery, 3
Late Period, 9
Late Pleistocene era, 135
Lavachery, Henri, 120
Lee, Vince, 82–83, 86
Leprosy, 174
Lisjanskij, Yuri, 153
Lithic mulching, 43–48, 50, 51, 191–197
Long Ears, 10, 50
Lord Howe Island, 28–29
Louis XVI, King of France, 37
Louwagie, Geertrui, 48, 190, 193–194
Love, Charlie, 62–69, 81, 86
L'univers (journal), 165

Machu Picchu, Peru, 82, 115
MacIntyre, Ferren, 80
Magnesium, 196–198
Mahine (translator), 160, 162
Maitaki Te Moa, Easter Island, 196
Mana (concept of supernatural power), 56
Manavai (rock circles), 38–41, 44, 50, 53, 123, 126, 140, 191–198
Mangareva Island, 167–169
Mangototo (statue), 162
Mann, Daniel, 23, 24
Maroowahai (islander), 160
Marquesas Islands, 2, 4, 9, 17, 120, 166, 167, 182
Marriage practices, 127, 128
Martinsson-Wallin, Helene, 112
Más Afuera Island, 163
Mata'a (tools), 95–99, 104
Mataveri, Easter Island, 173
McCall, Grant, 30
McCoy, Patrick, 19, 123
McCulloch, Allan, 28
Measles, 156
Merlet, Enrique, 170, 173
Métraux, Alfred, 120, 147
Middle Period, 9
Mieth, Andreas, 23, 24, 46
Miro-o-one (earth ship), 154, 155
Missionaries, 168–169
Moai. *See* Statues (*moai*) of Easter Island
Moerenhout, Jacques Antoine, 120
Mohican (ship), 171
Moore, Omar Khayyam, 137–139
Morrison, Alex, 191
Motu (islets), 95–96
Mounds, 122, 135–136
Mule deer, 100–101
Mulloy, William, 9, 20, 78
Mystery of Easter Island, The (Routledge), 55, 60, 73, 131

Nancy (ship), 163–164
NASA (National Aeronautics and Space Administration), 174
Naskapi Native Americans, 136–139
National University of Ireland, 80
Native Americans, 136–139, 157
Necker Island, 37–38
Negev Desert, Israel, 43
Neke-neke, 80
New Guinea, 106
New Mexico, lithic mulching in, 43–44
New Zealand, 4, 7, 17, 28, 161, 165
Nihoa Island, 38
Nitrogen, 47–48
Northern Line Islands, 38
Nucleated settlement pattern, 121, 123
Nuku Hiva Island, 120

Oahu, Hawaiian Islands, 25–27
Obsidian, 15, 42, 95–97, 99, 122, 181
Ollantaytambo, Peru, 115
Oral history, 9–10, 13, 60
Ordy Pond, Oahu, 26, 27
Orlebar, John, 164
Orliac, Catherine, 23, 34
Orliac, Michel, 34
Orongo, Easter Island, 171
Owls, 29

Paina (reed figures), 118–119
Pakomio, Leonardo Haoa, 80
Palmer, J. Linton, 21, 153
Paper mulberry, 38
Paro (statue), 143
Patoo-patoos (clubs)
Paumotu language, 168
Pavel, Pavel, 78–81, 86
Pearthree, Erik, 34
Peru, 82, 115, 165–166
Phoenix Islands, 38
Phosphorus, 38, 47, 190, 196–198
Photosynthesis, 48
Pickersgill, Lieutenant, 161
Pigs, 5, 34, 170
Pirca walls, 41, 170
Pitcairn Island, 4, 37, 178
  population of, 183–184
  stone statues on, 120
Plaggen soils. *See* Lithic mulching

INDEX

Plague, 156
Plantains, 34, 42
Plants. *See also* Soil; *specific trees*
bananas, 5, 21, 33, 35, 38, 40, 42, 160, 163, 183, 187, 188, 193, 194
breadfruit, 5, 182
kava, 5
plantains, 34, 42
potatoes, 21, 33, 42
sugarcane, 5, 21, 33, 34, 38, 42, 183, 187, 193, 194
sweet potatoes, 4, 34, 35, 42, 59, 163, 183, 190, 193, 194, 196
taro, 5, 6, 34, 35, 38, 40, 43, 46, 178, 182, 183, 187, 188, 194, 196
yams, 5, 34, 35, 42, 163, 183, 187, 194, 196
Platforms (*ahu*), 1, 13, 65, 70, 82–83, 86–87, 120. *See also* Statues (*moai*) of Easter Island
construction of, 111–113, 115, 117, 126
inventory of, 111–112, 171
names of, 111
rituals held at, 117–118
shape of, 90
spacing of, 128
Poe poe (rock piles), 192
Poike Ditch, Easter Island, 99
Poike Peninsula, Easter Island, 24, 52, 56
Poike volcano, Easter Island, 41, 50–52, 99, 181
Polet, Caroline, 142
Pollen evidence, 17, 19, 22–23, 26
Polynesia, 1–3. *See also specific countries and islands*
colonization of, 3–4
size of territory, 4
Pompeii, 73
Population, of Easter Island, 6, 8, 11, 24, 32, 53, 60, 93–94, 105, 111, 141–143, 158, 167, 170–171, 174–175, 183
Poro stones, 65, 116, 117, 123, 125
Post holes, 66–67
Posture, in face of threat, 133
Potassium, 38, 190, 196–198
Potatoes, 21, 33, 42
Price, George, 100–101
Pritchardia palm, 26, 28
Pukao hat forms, 126, 143, 151–152
Puna Pau, Easter Island, 151–152, 181

Radiocarbon dating, 9, 15–16, 22–23
Rainfall, 37, 47, 48, 181, 184–187, 189
Raivavae Island, 120
Randomization practice, scapulimancy as, 138–139
Rano Aroi, Easter Island, 22, 47, 181
Rano Kau, Easter Island, 22, 42, 47, 50, 56, 69, 95, 181
Rano Raraku, Easter Island, 22, 23, 55, 59–61, 63, 68, 69, 73–75, 77, 81, 86, 181
Rapa Iti Island, 99, 178
Rapa Nui. *See* Easter Island
Rapu, Alfonso, 175
Rapu, Sergio, 51, 62, 83, 85–87, 89, 117, 175
Raro-tonga Island, 4
Raroia Island, 9
Rats), 5
bird extinctions and, 28–29
in Chile, 30–31
on Easter Island, 29–31, 34, 35, 49, 51
on Hawaiian Islands, 27–28
on Lord Howe Island and, 28–29
reproduction of, 28, 29–30
Red scoria gravel, 116, 117, 126, 151, 152
Reproductive strategy, 134, 136, 139–140, 183
Requiem for a Nun (Faulkner), 177
Resolution (ship), 7, 161
Riddle of the Pacific, The (Brown), 33, 109
Roads, 56–71, 77, 83, 84, 92

Rock circles (*manavai*), 38–41, 44, 50, 53, 123, 126, 140, 191–198
Roggeveen, Jacob, expedition to Easter Island (1722) by, 6, 10, 21, 33–34, 36, 98, 118, 123–124, 147, 148, 155, 158
Root molds, 15, 20, 23
Rosendaal, Roelof, 147
Roussel, Hippolyte, 168–169
Routledge, Katherine, 95, 120, 172–173
  disease and, 174
  miro-o-one (earth ship) celebration, description of, 154–155
  The Mystery of Easter Island by, 55, 60, 73, 131
  oral history collected by, 9–10
  roads studied by, 59–63, 67, 68
  statues (*moai*) studied by, 75, 77, 84, 118, 119
Routledge, Scoresby, 172, 173
Russian expedition to Easter Island (1816), 164

Sagan, Carl, 177
Salmon, Alexander, 111
"Sam" (model), 81
Samoa, 3, 106
San Lorenzo (ship), 118
Santa Rosalia (ship), 96, 118, 152
Satellite images of Easter Island, 39, 40, 51, 67–70
Scapulimancy, 137–139
Science (journal), 17
Scurvy, 7
Sealers, 163
Seawall construction, 112–115, 126
Seeds, as rat food, 27–29
Selling, Olaf, 19
Seringapatam (ship), 153, 164
Settlement patterns, 121–123
  on Easter Island, 123–129
Sexual encounters with visitors, 159, 161, 163, 164, 169
Sheep, 31, 41, 46, 47, 52, 58, 59, 67, 169–174
Shifting cultivation, 49–50
Short Ears, 10, 50
Skeletal remains, on Easter Island, 94, 127–128, 141, 142
Skottsberg, Carl, 3
Slash-and-burn cultivation, 25, 27, 49–50, 53
Slave raids, 60, 106, 163–167
Smallpox, 156, 166, 167
Smith, John Maynard, 100–101, 145
Sober, Elliott, 145
Society Islands, 4, 120
Soil, 179, 189. *See also* Plants
  erosion, 23, 46, 47, 51, 52, 65, 66, 170, 190
  lithic mulching and, 43–48, 50, 51, 191–197
  manavai (rock circles) and, 38–41, 44, 50, 53, 123, 126, 140, 191–198
  mineral content of, 38, 47–48, 51, 190, 194–198
Sophora toromiro (bush), 34
South America, 4, 110, 114, 115. *See also specific countries*
South Pacific, currents of, 4
Southern Cook Islands, 5
Spanish expedition to Easter Island (1770), 6, 10, 34, 96, 99, 106, 118, 152, 158–159, 161, 162
Statues (*moai*) of Easter Island. *See also* Easter Island
  abandonment of construction of, 55, 77, 78, 83–85, 162
  benefits of making, 131–146
  carving and construction of, 73, 75, 77, 83, 86–87, 143
  center of mass (center of gravity) of, 87–91
  conchoidal fracture of, 85–86
  as costly signaling, 132–134, 136
  experiments on, 78–83, 86
  eye sockets of, 83
  heads of, 75, 77
  Heyerdahl's theory on, 110, 111, 115
  inventory of, 60, 171

"new," 69
Nova, 2000 television program about aired on, 82
number of, 2
partially completed, 73, 75
platforms of. *See* Platforms (*ahu*)
positioning of, 1, 2
shape of, 86–87, 90–91
size of, 1, 143, 162
temporary versions of, 118–119
theories of, 6, 7–12
transportation of, 11, 20, 53–54, 59–71, 73, 77–92
Stefan, Vincent, 128, 141
Stevenson, Chris, 44, 47, 191
Stonehenge, 121
Strakonice, Czechoslovakia, 79
Sub-Antarctic islands, 4
Sugarcane (*Saccahrum officinarum*), 5, 21, 33, 34, 38, 42, 183, 187, 193, 194
Suicides, 169
Supply (ship), 28
Surface rocks, soil under, 43, 44
Survival of the fittest, 134
Sweet potato (*Ipomoea batatas*), 4, 34, 35, 42, 59, 163, 183, 190, 193, 194, 196
Syphilis, 163

Tahiti, 2, 4, 17, 93, 166, 167, 169, 171
Taro (*Colocasia esulenta*), 5, 6, 34, 35, 38, 40, 43, 46, 178, 182, 183, 187, 188, 194, 196
Te Niu, Easter Island, 42–43, 46
Terevaka volcano, Easter Island, 47, 50, 51, 69, 99, 151, 181–182
Terra Australis, 7
Terracing, 35
Theory of the Leisure Class, The (Veblen), 133
Thienhoven (ship), 98, 147, 148, 155
Thomson, William J., 111, 171–172
Ti leaf, 34
Tonga, 3, 142
Topaze (ship), 147, 153
Toromiro (bush), 171
Tourism, 174, 175, 179
Trade winds, 4, 5
Transpiration, 185
Tschochner, Bernd, 46
Tuamotu Islands, 4, 9, 120, 155, 166–168
Tuberculosis, 156, 166, 169
Turmeric, 5
Twain, Mark, 100

U.S. Air Force, 174

Vaihu, Easter Island, 59
Valparaiso, Chile, 168
Van Tilburg, Jo Anne, 77, 81–82, 84, 86
Vanuatu Island, 154
Veblen, Thorstein, 133
Vegetables. *See* Plants
Venereal disease, 156, 158, 159, 162–163
Vitousek, Peter, 47

Waldegrave, Captain, 164
Washington Island, 38
Water, on Easter Island, 3, 6, 162, 175, 181–182
Weapons, 94–99, 104, 161
Westerly winds, 5
Whalers, 163, 164
Williamson, Balfour & Company, 170, 172–174
Wilson, David Sloan, 145
Windbreaks, evaporation and, 35
Winds, 4–5, 187–189
Wofford, Gabe, 196
Wood charcoal, 16
Wozniak, Joan, 42–45

Yam (*Dioscorea alata*), 5, 34, 35, 42, 163, 183, 187, 194, 196
Yucca, 34

About the Authors

Terry Hunt has spent the past thirty-five years doing archaeological field research in the Hawaiian Islands, New Zealand, Fiji, Samoa, Papua New Guinea, and Rapa Nui. He received his doctorate at the University of Washington. Currently professor of anthropology at the University of Hawaii at Manoa, he has been honored with the University of Hawaii Board of Regents Medal for Excellence in Research for both his teaching and his research efforts at Easter Island and elsewhere. He lives in Hawaii. You can visit his website at www.anthropology.hawaii.edu/People/Faculty/Hunt/.

Carl Lipo is an associate professor of anthropology at California State University Long Beach (CSULB). He received his PhD from the University of Washington and began working in collaboration with Terry Hunt on Rapa Nui in 2003. In 2010 he was awarded the Distinguished Faculty Scholar and Creative Achievement Award by CSULB. He lives in Long Beach, California. You can visit his website at www.lipolab.org.